PRAGMATICS AND EDUCATION

LANGUAGE AND LANGUAGE ACQUISITION
Series Editor: **F. Lowenthal,** *Mons University, Mons, Belgium*

LANGUAGE AND LANGUAGE ACQUISITION
Edited by F. Lowenthal, F. Vandamme, and J. Cordier

PRAGMATICS AND EDUCATION
Edited by F. Lowenthal and F. Vandamme

A Continuation Order Plan is available for this series. A continuation order will bring delivery of each new volume immediately upon publication. Volumes are billed only upon actual shipment. For further information please contact the publisher.

PRAGMATICS AND EDUCATION

EDITED BY

F. LOWENTHAL

Mons University
Mons, Belgium

AND

F. VANDAMME

Ghent University
Ghent, Belgium

PLENUM PRESS • NEW YORK AND LONDON

Library of Congress Cataloging in Publication Data

Pragmatics and education.

"Based on the results arising from the Third Language and Language Acquisition Conference on Pragmatics and Education, held March 21-25, 1983, in Ghent, Belgium" —T.p. verso.

Includes bibliographies and index.

1. Pragmatics. 2. Language and languages—Study and teaching. 3. Language acquisition. 4. Cognition in children. 5. Language disorders in children. 6. Mathematics—Study and teaching. I. Lowenthal, F. II. Vandamme, Fernand J. III. Colloque "Langage & acquisition du langage" (3rd: 1983: Ghent, Belgium)

P99.4.P72P738 1986 418'.007'1 86-91549

ISBN 0-306-42374-X

Based on the results arising from the Third Language and Language Acquisition Conference on Pragmatics and Education, held March 21-25, 1983, in Ghent, Belgium

A Division of Plenum Publishing Corporation
233 Spring Street, New York, N.Y. 10013

Printed in the United States of America

INTRODUCTION

F. Lowenthal
Universite de l'Etat a Mons
24 rue des Dominicains
7000 Mons
Belgium

The series of "Language and Language Acquisition" conferences was born in Mons in 1977. One day the Dean said to me: "You are doing research in that field, why don't you try to organize a small conference?". I thought about it, tried to contact people, received several answers and finally told the Dean: "There will be so many participants and I need so much money to organize the conference". His answer was a short one: "I told you to organize a SMALL conference". I do not know what he did, but he succeeded in working a miracle: the funds were found and the conference took place. This miracle has been repeated twice: once in Mons (1980) and once in Ghent (1983). The group of people interested in these conferences has become bigger, but the aim of the organizers is still the same: to bring together people working in different fields such as mathematics, philosophy, linguistics, logic, computer, science, education, psychology, medicine, ... and to give them the possibility to have long discussions even if the time devoted to the presentation of papers has to be reduced.

This book does not contain the proceedings of any of the conferences - they have been published separately - but the international committee of the language and language acquisition group thought that it would be interesting to produce a book containing a set of selected papers which all concern different but complementary approaches of "Pragmatics and Education". The approaches illustrated in this book are the following: educational and cognitive aspects of discourse, non-verbal communication devices, mathematics education, native (and non-native) language acquisition and learning, and language disorders. The book also contains a short synthesis of the discussion sessions.

All the papers were written after the conference by selected researchers; they were asked to write detailed papers which were not only based on their own experience, on their personal experiments, but also on everything they heard during the conference and especially during the discussion sessions: some of the papers concern research which only started after the conference was over but was inspired by the discussions.

I wish to thank here all those who made this book possible: the Universities who helped us to organize the conference, my colleagues in Ghent who took care of the local organization, the participants for accepting to share their ideas, the authors for taking these ideas into account when they wrote their special paper. I also wish to thank all those who worked very hard, but very discreetly, to type all these papers.

CONTENTS

DISCOURSE EDUCATION AND COGNITIVE ASPECT

PRAGMATICS AND LANGUAGE DIFFERENCES

Eric Esperet

Laboratoire de Psychologie
E.R.A. CNRS 797
Université de Poitiers, France

Pupils' linguistic performances are often mentioned to explain academic failure. Moreover, they are supposed to differentiate children according to their social origin. So an attempt has been made for many years to describe these language variations and to analyse the mechanism of their genesis. This topic has formed the subject of several successive formulations, particularly in the field of differential psychology. But the different analyses thus proposed have raised many theoretical and methodological problems. These problems will be examined in the first part of this study, in order to determine which questions have yet to be answered satisfactorily. Then we shall try to specify the form of the pragmatic approach in psycholinguistics, and what this trend brings to the study of language differences: does it allow, for example, the discovery of more responses, or at least the suggestion of more heuristic formulations to an old problem? The third part will deal with several studies conducted by our team, which will illustrate the trend of this paper.

I. LANGUAGE DIFFERENCES: THE EVOLUTION OF PYSCHOLOGICAL CONCEPTIONS

The study of language differences is both old and new: from the very beginning of this century, the large variations in the scores obtained in verbal tests by different people or different social groups have been noticed. But only in the last decade has a differential analysis of situational discourse been developed. The review of psychological litterature thus leads us to distinguish three successive stages, of unequal duration, which correspond each to a dominant formulation (Espéret, 1982b):

1. The first approach is essentially psychometric and was used up to the beginning of the 1960s. It is characterized by the use of standardized situations, in which verbal elements (phonemes, words or sentences) must be processed by the subject. From the scores thus obtained, the studies of different factorialists, such as Thurstone, Thomson or Burt, have arrived at the notion of verbal ability for wich people were differentiated according to their sex, social origin or their intellectual level.

This first trend remains rather isolated from other psychological fields: verbal learning theories, mostly behaviourist, rarely consider differential aspects (cf. Noble, 1961); the developmental study of language only mentions them secondarily (cf. Buhler, 1931; Davis, 1937; or Smith, 1935). Finally, let us recall that psycholinguistics "stricto sensu" only appears towards the end of this period.

2. The second stage clearly corresponds to a sociolinguistic dominant. The change of viewpoint, from the 1960s onwards is mainly due to the influence of trends exterior to psychology: ethnography of communication (Gumperz, 1962; Hymes, 1964); social linguistics (Labov, 1966) and sociology (Bernstein, 1959, 1960; Bourdieu and Passeron, 1964). This infatuation of psychologists for sociolinguistic analysis came frome the growing negative reactions to both the notion of the "ideal speaker" and the classic psychometric approach. The reference to Bernstein's theory has thus become a "standard": it indeed constituted the first attempt to analyse the intermediate processes between belonging to a social class and the use of language during socialization.

For about 15 years, numerous studies have thus been carried out within this framework. They have constituted a practically autonomous field of research: for example, interactions with social psychology or developmental psycholinguistics have remained rare.

3. The third stage began in the mid-1970s. It possesses two characteristics: the non-differentialists'growing interest for individual variations in language; the disappearance of a unitary dominant conception in the study of language differences. The search for a global theory becomes less important, and social variations are approached in a specific way through four psychologival areas: differential psychology which begins to pay attention to the representativeness of the standardized situations used, and to dimensions of language other than formal ones (Carroll, 1979); the general psychology of language which tries to integrate differential aspects to the analysis of information processing (Hunt et al., 1975); developmental psycholinguistics for which interindividual variations belong to acquisition processes (Nelson, 1973; Nelson, 1981); and finally social psychology which, often in a school context, has determinedly tackled the study of the inter-

individual processes underlying the social variations of language (the "Bristol group" is an outstanding example). For three of these sectors pragmatic aspects play an important role in language analysis. Does this intervention of pragmatics permit light to be shed on the problem of linguistic differences? Before answering, it is necessary to specify the points yet unanswered by the first two stages described above.

The study of language differences, particularly the ones linked to social origin, tries to answer two main questions: 1. Do these differences exist and in what conditions? 2. Which theoretical analysis of their genesis and their consequency can we suggest? For each of these questions, the answers found by the first two stages remain very incomplete:

1. Till the 1950s, descriptive results were limited but fairly coherent: in standardized situations children of the lowest socio-economic status generally obtained the lowest scores, in terms of raw scores, verbal factor scores, or even of verbal intelligence. Furthermore, most of the results were established within the framework of structural studies of intelligence and not language studies: Anastasi's book (1958), long considered to be the reference in differential psychology, quotes only two empirical studies concerning corpora, and for the rest refers to Mc Carthy's chapter, in Carmichael's famous textbook (Mc Carthy, 1952).

In the sociolinguistic stage, the situations used are more varied and this entails a real decrease in the coherence of results. Without going into all the details (cf. Espéret, 1979 and 1982a), we realize that classically admitted differences vary in amplitude or disappear when the research situations, levels of language analysis or even individual characteristics within the same social group, are changed. These variations concern not only the formal aspects of language (lexicon, morphosyntactic features) but also functional aspects (cf. research on referential communication, following Heider, 1972, or the ones prompted by the two codes described by Bernstein). Many criticisms have underlined the gaps in these descriptive studies: the speaker's perception of the situation is not taken into account; the communicative function of language and its role in the psychological regulation of social interactions is neglected; the notion of code, which would characterize a social group, remains an artificial construction without any steady empirical content. The previous certainties have disappeared; it is necessary to reexamine the existence of language differences but on new bases.

2. The basic reason for contradictory results seems to come from the theoretical gaps in research from a psychological viewpoint. The psychometric studies had only fairly superficial hypotheses for the genesis of language differences: few verbal

stimulations, a poor parental linguistic model, etc. Concepts such as verbal ability and verbal intelligence, in factor analysis, are often used as ultimate explanations for bad verbal performances. But the tautological character of this explanatory schema has not permitted advancement in the theoretical elaboration of the problem (Espéret, 1980). As for the sociolinguistic stage, it is necessary to reconsider the criticism put forward by Wells, Giles and Robinson (Wells and Robinson, 1982; Giles et al., 1980): the maze of empirical research has gradually been organized under the influence of Bernstein's conceptualizations (between 1959 and 1970; cf.Bernstein, 1971). But this attempt, although coherent and important on the sociological level, has always suffered from an almost total absence of psychological theory concerning both language acquisition and its relations with cognitive development. From an operational viewpoint, therefore, the question remains unanswered: what aspects of language necessitate research into interindividual differences? From a theoretical viewpoint, it is more than ever necessary to have a model of language acquisition and functioning which permits the justification and interpretation of the selected aspects. This model did not yet exist at the beginning of the 1970s, a fact which is partly explained by the rather recent advances in developmental psycholinguistics. But this absence has made any real improvement in the field examined here difficult.

II. THE PRAGMATIC APPROACH TO LANGUAGE DIFFERENCES

Can pragmatics supply this model or contribute to its construction? To answer this question, it is first necessary to examine the contribution of pragmatics to the psychology of language; then we shall see if this contribution can create a new theoretical analysis of language differences.

A. Pragmatics and psycholinguistics

Psychological pragmatics does not constitute a coherent and structured field. Rather it is defined by common preoccupations well resumed by Markova (1978) or Ervin-Tripp and Mitchell-Kernan (1977); the analysis of the role played by context in the comprehension and production of utterances; the functional approach of language; work on verbal interaction in defined situations, particularly on conversations; and sometimes taking into account the suprasentence level in discourse processing. These preoccupations appear in three main research topics:

1. The acquisition of direct or indirect speech acts: these studies are prompted by Austin's (1962) and Searle's (1969,1975) conceptions. They point out how the child learns how to master linguistic forms and social origin conventions which regulate the use of utterances at the same time. Emphasis is thus laid on

social acts, linguistically performed, in specific contextual conditions. The analysis of these speech acts, particularly during the prelinguistic period, raised problems of taxonomy for which several solutions were suggested (Bates et al., 1975; Bruner, 1975; Dore, 1975, 1977, 1978). Furthermore, the process which would enable one to go from the structure of communicative intention to the linguistic structure of the utterance still gives rise to fairly divergent conceptions (cf. the controversy between Bruner and Dore, in the book edited by Markova, 1978). The most radical position is developed by Halliday (1975, 1978): the form taken by an utterance is a direct result of the functions it fulfills.

Nevertheless, this research field clearly emphasizes the need to analyse the functions of the utterances gradually mastered by the child.

2. Learning communicative behaviours: well illustrated by Bruner's works (1975, 1978, 1980, 1981), these studies show how verbal exchanges between the child and his environment take root from non-verbal interactions which exist before them and prepare them, the study of look exchanges, of turn-taking in children beginning to speak, of the establishing of conventional "formats", emphasizes that dialogue begins before language and even that the latter is created from these exchanges. Progressively a "conversational competence" also emerges: the child becomes able to monitor his interventions by taking into account his interlocutors and minimal rules of co-operation (cf. Ochs and Schieffelin, 1979; Ervin-Trip and Mitchell-Kernan, 1977; and Martlew, 1980). The setting-up of this ability to converse efficiently seems to depend closely on the way the environment interacts with the child; in particular, as far as the step negotiation of shared meanings is concerned (Wells, 1980; Wels and Robinson, 1982). The acquisition of linguistic forms would then depend more on the parental will to maintain an efficient communication than on didactic corrections of the child's first utterances.

The idea that a "degenerate" linguistic input is presented to the child has been widely disproved (Snow and Ferguson, 1977). But the environment's adaptation to the young child might involve a permanent monitoring of the dialogue, such as the one described by Bruner (1981) as a "language assistance system" focussing on a gradual transfer of initiative.

3. The third topic concerns older children and adults. It analyses how world knowledge possessed by a speaker intervenes in the processing applied to an utterance. This is called pragmatic strategy, as opposed to morphosyntactic strategy; for example, in the processing of reversible or irreversible passive sentences (Bronckart et al., 1976). Similarly, the phenomenon of presupposition and inference (Dore, 1976; Kail, 1978) calls for socially

acquired knowledge as well as well as syntactic decoding. Previous studies were generally limited to the utterance level. However propositions exist to extend these processes to the discourse level, with cognitive text processing in mind (Freedle and Duran, 1979; Freedle and Fine, 1982; Van Dijk, 1980a, 1982); the numerous studies on narrative are a good illustration. Finally it is necessary to add that metacognitive structure of pragmatic origin can intervene in comprehension and production processes (Flavell, 1981; Markman, 1981). The rapid examination of these three topics allows us to specify those pragmatic aspects which could affect the study of language differences:

- Uniquely formal analyses of language (syntax, for example) give way to semantic and functional analyses.
- The roots of language are looked for at ever earlier ages, particularly in preverbal communication.
- Linguistic development is often analysed through actual verbal interactions and not through single utterances.
- Finally, this development is related to cognitive development in a more explicit way (cf. Bates, 1976; Karmiloff-Smith, 1979).

It remains to be seen if, pragmatics has prompted a change in differential psychology studies on these different aspects.

B. Contributions of pragmatics to the study of language differences

Individual variations are frequently taken into account in the field of pragmatics. Even if the studies mainly focussed on this question are not yet numerous, there have already been important contributions:

1. The notion of individual strategy has aroused more and more interest in psychologists who supported the idea that acquisition did not necessarily take place in the same way for everybody. The excellent review made by Nelson (1981) describes some of its mechanisms; most of the studies mentioned differentiate two great functional strategies in language use: referential expression and the mastery of interpersonal relations (cf. also mathetic and pragmatic functions, in Halliday); recent studies specify the modes of linguistic realization. A child's "choice" of strategy would then come from the emphasis placed upon a particular function through different contexts by the environment; the "forms" preferentially implemented would issue from the type of social interaction experienced by the child. Here is an idea already put forward by Bernstein, but little explored empirically, according to which the language used would correspond to the perception of reality of the person concerned. The existence of social differences in functional preferences (Tough, 1981; and for question processing: Gullo, 1981; Simmons, 1976), prompts strategies noticed previously to be otherwise re-formulated; it also permits the partial analysis of the role played by context through the representations built up by the child.

From this point of view, the absolute comparison between children from different social origins, on a verbal-nominal dimension for example, is no longer relevant. It can only correspond to different strategies and not to hierarchical levels in language development.

2. The most important contribution lies in the accurate analysis of interactions between the child and his environment (cf. the partial review presented by Rondal, 1983). This field of investigation has greatly developed in 10 years and has produced the empirical studies which were missing on this point.

Several noticeable precisions have been made on the cognitive and social processes which influence language development: the type of mother-child interaction seems to play a greater role than social class (Borduin and Henggeler, 1981; Adams and Ramey, 1981; Wells, 1980); this interaction presents large interindividual variations, which in some aspects can covary with the social origin (Farran and Haskins, 1980; Farran and Ramey, 1980).

More important from a theoretical point of view, the central mechanism of interaction seems to be the continuous negotiation, engaged between mother and child, to maintain a meaningful communication; this negotiation concerns both the semantic content (the "tuning" in for a shared topic) and the conversational rules (speech turns for example); moreover, through this process, the mother is also influenced by her child; the adaptation, therefore, is reciprocal (Borduin and Henggeler, 1981; Lieven, 1978). So what differentiates the mothers of different social origin is their sensitivity to the child's needs and abilities in the interaction (Ninio, 1980). With all these aspects, a satisfactory psychological approach can be inspired by Bruner's analyses (ibid.); in particular through the process he describes, allowing the child to withdraw the elements of the action which are the precursors of grammatical cases (Bruner, 1978).

However one problem remains unsettled: the nature of the psycholinguistic mechanisms which make it possible to go from communicative intention (function of utterance) to the morphosyntactic structure (linguistic form of utterance). The second analysis of the elaborated code, made by Young (1982) in the framework of Halliday's conceptions, for example, does not appear very convincing on this point.

Maybe it is more efficient to try to tackle the problem through a model of the speaker which calls upon notions such as attitudes and representations (Fielding and Fraser's attempts, 1978; or the analysis of left-dislocation in Italian by Ochs, 1979).

III. TOWARDS A DIFFERENTIAL ANALYSIS OF PRODUCTION PROCESSES

Voluntarily, the analysis presented here has been limited to the central cognitive mechanisms. It would also be necessary to take into account the studies carried out, for example, in the school situation, and centered on the forms of exchange and the psychosocial factors regulating communication (cf. Stubbs and Delamont, 1976; Sinclair and Coulthard, 1975). But this restriction results from the conviction that, in the field of linguistic differences, the main questions are situated at this level of psychological analysis.

A. Pragmatics and the construction of a model of language acquisition

Does pragmatics provide a theoretical model of language acquisition at present? The answer is certainly negative. On the other hand, this trend has widely influenced the way in which language acquisition is studied and the integration of interindividual differences to the studied processes. Therefore differential psychologists have only to work on two points mentioned in this contribution: 1. To analyse the role played by representations elicited in the different interlocution situations in language construction; if we want to integrate the elements of context to language processing mechanisms, they must be represented as cognitive schemata, monitoring the operations of coding and decoding utterances in interaction with the processes of morphosyntactic processing. Then we should try to differentiate the speakers, not according to their productions, but according to the production processes themselves (partially at least) and thus formulate hypotheses which can be checked empirically. 2. To re-analyse the different acquisition processes, including the different kinds of learning (Moerk, 1984): the aim would no longer be to build up a unitary theory of language development, which is probably nothing more than a convenient myth. We should try to analyse the way imitation, conditioning and cognitive processes intervene specifically at different levels of language, and for different types of acquisition (lexicon, syntax, communicative competence, particular textual category, etc.).

A thorough study of these two directions would next allow the problem of linguistic differences linked to social origin to be brought up again, this time differently and supported by psychological arguments. The significant decrease, in the last seven years, of the number of references to a theory such as Bernstein's, no doubt reflects the need felt for a better foundation of hypothesized psychological mechanisms of social environment effects.

More generally, the study of language differences must then present the following characteristics:

- To restrict the approach to a specific language behaviour, which can be functionally characterized (i.e. situation of implementation, finality, involved psycholinguistic operations).

- To have, as much as possible, a general model of this particular behaviour which allows expression of the speaker's functioning in the form of fundamental processes leading to the production of discourse in a situation.
- To analyse language differences in terms of variation of these fundamental processes (speaker's functioning), and not in terms of categorized productions (code, for example).
- To look for individual or social characteristics which can affect the production processes themselves, in their genesis or their functioning.

B. Differential analysis of narrative behaviour in a school situation

In this perspective, our team has been carrying out research on the construction of the narrative behavior in children between the ages of 4 and 11 years for the last 3 years. This behavior was chosen because it provides a framework already well studied and plays an important role in school adaptation (Feagans, 1982). We started from a theoretical model of narrative production (Espéret and Gaonac'h, 1983; Espéret, 1984) which includes genetic and differential aspects. This model finds a great part of its inspiration in theories of cognitive schema (cf. Labov and Waletzky, 1966; Rumelhart, 1975; van Dijk, 1980b) and so refers to the different representations of this behavior built up gradually by the child.

These representations are supposed here to assure a monitoring role on the operations of selection of the content (topic choice; construction of macro and micro-structures) and on psycholinguistic operations of wording (enunciative and temporal marks, coherence) in storytelling. One of our objectives is to analyse why children produce different narratives, from the points of view of language use and content organization. We try to show that these variations in productions are due to differences in the mastery of implied processes (the degree of possession of the narrative schema for example) and that these differences explain those, more common, observed at the language level. Some results can partially illustrate this approach.

1. We were able to verify that between 5 and 11 years narrative schema is progressively built up (Espéret and Chauvineau, 1981; Ballaire, 1982). This occurs in several stages: mastery of textual coherence, then representation of the narrative event and finally integration of the narrative event into a more general schema: initial situation - complication - solution - conclusion. The order of the stages seem to be more or less constant but the level reached markedly differentiates children of the same age. In this case, the narrative schema can be compared to a metacognitive structure.

2. Parallel to the construction of the narrative schema, children are more and more able to produce imaginary narratives which translate this schema into a text (Espéret, 1984). This translation can be seen in the organization of the content according to the narrative categories (initial situation, complication, etc.) and by language marks specific to narratives (introducers or verbal tenses: cf. Gaonac'h and Espéret, 1984).

Furthermore, we have shown children who knew the narrative schema well could better understand and memorize a narrative text than children of the same age who didn't know it (Espéret and Chauvineau, 1984).

We are presently carrying on this work, to specify which role socially acquired cognitive structures, play during on-line production. Finally, the influence exercised by some family practices (narrative reading, kind of parent - children discussions) on this structure is also studied (cf. also Cession et al., 1984).

Taken as a whole, it therefore appears more interesting from a theoretical point of view to look for the mechanisms which give rise to different language productions than to describe the actual productions. Differences at this level of discourse monitoring will be interpreted and taken into account in further testable hypotheses. Then pragmatics will make the analysis of the mechanisms supposedly present possible; the role played by the semantic structuration is rightly emphasized at the text level, and not at the utterance level; finally pragmatics will specify how the processes of discourse monitoring are elaborated through child-environment interaction.

However, it is now up to differential psychologists to propose models, limited to specific behaviors, which account for the observed language variations. We can think of educational interventions which modify the actual sources of language differences - and not only the differences (Chauvineau and Espéret, 1984).

REFERENCES

Adams, J.L., Ramey, C.T., 1980, Structural aspects of maternal speech to infants reared in poverty. Child development, 51, 1280-1284.

Anastasi, A., 1958, Differential Psychology. New York, Mc Millan Company.

Austin, J.L., 1962, How to do things with words. Cambridge (Mass.), Harvard Univ. Press.

Ballaire, M.F., 1982, L'événement narratif. Roneoted paper, Laboratoire de Psychologie de Poitiers.

Bates, E., 1976, Language and context: studies in the acquisition of pragmatics. New York, Academic Press.

Bates, E., Camaioni, L., Volterra, V., 1979, The acquisition of performatives prior to speech (1975). Reprinted in E. Ochs & E. Schieffelin, eds., Developmental Pragmatics. New York, Academic Press.

Bernstein, B., 1959, A public language: some sociological implications of a linguistic form. Britisch Journal of Sociology, 10, 311-326.

Bernstein, B., 1960, Language and social class. Britisch Journal of Sociology, 11, 271-276.

Bernstein, B., 1971, Class, codes and control. Vol. 1 London, Routledge & Kegan Paul.

Borduin, C.M., Hengeller, S.W., 1981, Social class, experimental setting and task characteristics as determinants of mother-child interaction. Developmental Psychology, 17, 209-224.

Bourdieu, P., Passeron, J.C., 1964, Les héritiers. Paris, Editions de Minuit.

Bronckart, J.P., Sinclair, H., Papandropoulou, I., 1976, Sémantique et réalité psychologique. Bulletin de Psychologie, numéro spécial "La mémoire sémantique", 225-231.

Bruner, J.S., 1975, The ontogenesis of speech acts. Journal of Child Language, 2, 1-19.

Bruner, J.S., 1978, From communication to language: a psychological perspective. In I. Markova, ed., The social context of language. New York, Wiley.

Bruner, J.S.,1980, Afterword. In D.R. Olson, ed., The social foundations of language. New York, Norton and Cie.

Bruner, J.S., 1981, The pragmatics of acquisition. In W. Deutsch, ed., The child's construction of language. New York, Academic Press.

Buhler, C., 1931, Kindheit und Jugend. Leipzig, Hirzel.

Cession, A., Kilen, A., Denhiere, G., Rondal, J.A., 1983, "Maman... une histoire". Roneoted paper. University of Liège and University of Paris.

Carroll, J.B., 1979, Psychometric approaches to the study of language abilities. In C.J. Fillmore, D. Kempler & W.S.Y. Wang, eds., Individual differences in language ability and language behavior. New York, Academic Press.

Cazden, C.B., 1966, Subcultural differences in child language: an interdisciplinary review. Merrill-Palmer Quarterly, 12, 185-218.

Chauvineau, J., Espéret, E., 1984, Conduite de récit et approche des difficultés en lecture. Colloque international "Perspectives de Réussite", Bordeaux, 27-29, Février.

Davis, E.A., 1937, The development of linguistic skills in twins, singletons with siblings and only children from age five to ten years. Institute Child Welfare Monograph Service, nr. 14. Minneapolis University, Minnesota Press.

Dore, J., 1975, Holophrases, speech acts and language universals. Journal of Child Language, 2, 21-40.
Dore, J., 1976, Children's illocutionary acts. in R. Freedle, ed., Discourse relations: comprehension and production. Hillsdale N.J., Lawrence Erlbaum.
Dore, J., 1977, "Oh them sheriff": a pragmatic analysis of children's responses to questions. In S. Ervin-Tripp and C. Mitchell-Kernan, eds., Child discourse, New York, Academic Press.
Dore, J., 1978, Conditions for the acquisition of speech acts. In I. Markova, ed., The social context of language, New York, Wiley.
Ervin-Tripp, S., Mitchell-Kernan, C., 1977.,eds., Child discourse, New York, Academic Press.
Espéret, E., 1979, Langage et origine sociale des élèves. Berne, P. Lang.
Espéret, E., 1980, Intelligence verbale et milieu social. Bulletin de psychologie, 32, 475-485.
Espéret, E., Chauvineau, J., 1981, La genèse de la notion d'histoire. Roneoted paper. Laboratoire de psychologie de Poitiers.
Espéret, E., 1982a, Variations des conduites langagières: codes ou codages différents? Bulletin de la S.F.E.C.A., n° 2, 339 - 351, Actes du Colloque "Variabilité interindividuelle et plasticité du comportement".
Espéret, E., 1982b, Langage et école: renouvellement nécessaire de la problématique classique. Comptes-rendus des IVemes Journées de Psychologie Différentielle, Université de Lyon II.
Espéret, E., 1984, Processus de production: genèse et rôle du schéma narratif dans la conduite de récit. In M. Moscato & G. Pieraut Le Bonniec, eds., Ontogénèse des processus psycholinguistiques et leur actualisation, Paris, Presses Universitaires de France.
Espéret, E., Chauvineau, J., 1984, Conscious narrative schema in text comprehension and recall: a differential approach. International Journal of Behavioral Development, in press.
Espéret, E., Gaonac'h, D., 1983, What does storytelling mean for children: narrative schema representation and storytelling at different ages. In S.R. Mathews II, ed., Supplementary Proceedings of the International Symposium on Text Processing, E.R.D.C., University of West Florida.
Farran, D.C., Haskins, R., 1980, Reciprocal influence in the social interactions of mothers and three-years-old children from different socioeconomic backgrounds. Child development, 51 780-791.
Feagans, L., 1982, The development and importance of narratives for school adaptation. In L. Feagans & D.C. Farran, eds., The language of children reared in poverty. New York, Academic Press.
Fielding, R., Fraser, C., 1978, Language and interpersonal relations. In I. Markova, ed., The social context of language. New York, Wiley.

Flavell, J.H., 1981, Cognitive monitoring. In W.P. Dickson, ed., Children's oral communication skills. New York, Academic Press.

Freedle, R., Duran, R.P., 1979, Sociolinguistic approaches to dialogue with suggested applications to cognitive science. In R. Freedle, Ed., New directions in discourse processing. Norwood, Ablex Publishing Company.

Freedle, R., Fine, J., 1982, Prose comprehension in natural and experimental settings: the theory and its practical implications. In S. Rosenberg, ed., Handbook of applied psycholinguistics. Hillsdale, N.J., Lawrence Erlbaum.

Gaonac'h, D., Espéret, E., 1984, Fonction des temps verbaux dans la production de récits libres. Evolution génétique entre 4 et 8 ans. To appear in J. Wittwer, ed., La psycholinguistique textuelle. No spécial. Bulletin de Psychologie.

Giles, H., Smith, P.M., Robinson, W.P., 1980, Language: Social psychological perspectives. Oxford, Pergamon Press.

Gullo, D.F., 1981, Social class differences in preschool children's comprehension of Wh-questions. Child Development, 52, 736-740.

Gumperz, J.J., 1962, Types of linguistic communities. Anthropological Linguistics, 4, 28-40.

Halliday, M.A.K., 1973, Explorations in the functions of language. London, Edward Arnold.

Halliday, M.A.K., 1975, Learning how to mean. London, Edward Arnold.

Halliday, M.A.K., 1978, Language as social semiotics. London, Edward Arnold.

Heider, E.R., 1971, Style and accuracy of verbal communications within and between social classes. Journal of Personality and Social Psychology, 18, 33-47.

Hunt, E., Lunneborg, C., Lewis, J., 1975, What does it mean to be high verbal? Cognitive Psychology, 7, 194-227.

Hymes, D., 1964, Toward ethnographies of communication. American Anthropologist, 66, 12-25.

Kail, M., 1978, La compréhension des présuppositions ches l'enfant. Année Psychologique, 78, 425-444.

Karmiloff-Smith, A., 1979, A functional approach to child language, Cambridge, Cambridge Univ. Press.

Labov, W., 1966, The social stratification of speech in New York City. Washington, Center for applied linguistics.

Labov, W., 1976, Sociolinguistique. Paris Editions de Minuit.

Labov, W., Waletzky, J., 1966, Narrative analysis. Oral versions of personal experience. In J. Helm, ed., Essays on the verbal and visual arts. Seattle, Univ. of Washington Press.

Lieven, E.V.M, 1978, Conversations between mothers and young children: individual differences and their possible implications for the study of language learning. In N. Waterson & C. Snow, eds., The development of communication. New York, Wiley.

Markman, E.M., 1981, Comprehension monitoring. In W.P. Dickson, ed., Children's oral communication skills. New York, Academic press.

Markova, I., 1978, The social context of language. New York, Wiley. Introduction.

Martlew, M., 1980, Mother's control strategies in dyadic mother-child conversations. Journal of Psycholinguistic Research, 9, 327-347.

Mc Carthy, D., 1952, Le développement du langage. In L. Carmichael, Ed., Manuel de Psychologie de l'enfant. Paris, PUF, Tome II (Translation).

Moerk, E.L., 1984, A differential interactive analysis of language teaching and learning. To appear in Discourse Processing.

Nelson, K., 1973, Structure and stategy in learning to talk. Monographs of the Society for Research in Child Development, 38, n° 149.

Nelson, K., 1981, Individual differences in language development: implications for development and language. Developmental Psychology, 17, 170-187.

Ninio, A., 1980, Picture book reading in mother-infant dyads belonging to two subgroups in Israël. Child Development, 51, 587-590.

Noble, C.E., 1961, Verbal learning and individual differences. In C.N. Cofer, ed., Verbal learning and verbal behavior. New York, Mc Graw Hill.

Ochs, E., 1979, Social foundations of language. In R. Freedle, ed., New directions in discourse processing. Noorwood, Ablex Publishing Company.

Ochs, E., Schieffelin, B. eds., 1979, Developmental pragmatics. New York, Academic Press.

Rumelhart, D.F., 1975, Note on a schema for stories. In D. Bobrow & A. Collins, eds., Representation and understanding. New York, Academic Press.

Searle, J., 1969, Speech acts: an essay in the philosophy of language. London, Cambridge Univ. Press.

Searle, J., 1975, Indirect speech acts. In P. Cole & J.L. Morgan, eds., Syntax and semantics. Vol III. New York, Academic Press.

Simmons, B., 1976, A linguistic analysis of disadvantaged kindergarten children's verbal responses to questions. Journal of Educational Research, 69, 253-255.

Sinclair, J.M., Coulthard, R.M., 1975, Towards an analysis of discourse. Oxford, Oxford Univ. Press.

Smith, M.E., 1935, A study of some factors influencing the development of the sentence in the preschool children. Journal of Genetic Psychology, 46, 182-212.

Snow, C.E., Ferguson, C., eds., 1977, Talking to children. Cambridge, Cambridge Univ. Press.

Stubbs, M., Delamont, S., 1976, Explorations in class-room observation. New York, Wiley.

Tough, J., 1974, Children's use of language. Educational Review, 26, 166-179.

Van Dijk, T.A., 1980a, The semantics and pragmatics of functional coherence in discourse. In J. Boyd & A. Ferrara, eds., Speech acts: ten years after. Bologne, Bompiani.

Van Dijk, T.A., 1980b, Macrostructures. Hillsdale, N.J., Lawrence Erlbaum.

Van Dijk, T.A., 1982, Attitudes et compréhension de textes. Bulletin de psychologie, 35, n° Spécial "Langage et compréhension", 557-569.

Wells, G., 1980, Adjustements in adult-child conversations; some effects of interaction. In H. Giles, W.P. Robinson & P.M. Smith, eds., Language social psychological perspectives. Oxford, Pergamon Press.

Wells, G., Robinson, W.P., 1982, The role of adult speech in language development. In C. Fraser & K.R. Scherer, eds., Advances in the social psychology of language. Cambridge and Paris, Cambridge Univ. Press & Ed. M.S.H., 1982.

Young, G.M., 1982, The elaborated code: a new formulation within a functional framework. Language and Speech, 25, 81-93.

DISCURSIVE REPRESENTATION

Jean-Blaise Grize

Centre de Recherches Sémiologiques

Université de Neuchâtel, Suisse

INTRODUCTION

A distinction can usefully be drawn to clarify what follows between two types of language which both contribute - to a greater or a lesser extent - to the elaboration of the representations of knowledge. One of these types consists of the family of logico-mathematical languages and the other of natural languages. No doubt they have something in common, but they differ in two respects: the objects they deal with and their function. The first type of language makes it possible for us to calculate, and the second to communicate.

Let us consider the objects they deal with first. The objects defined by logico-mathematical languages are characterized by the univocity and the precision of their definitions. A "triangle" in Euclidian geometry, a "mass" in classical physics, an "element" in modern chemistry are what they are and it would be unthinkable to use them with another meaning. On the other hand, the objects of natural languages are always more or less indeterminate, and vague. It is normal, therefore, that their meaning depends on their context: for instance "mind" in the two expressions "mind the step" and "mind the children".

Now let's move on to the difference in function. It all depends on the room they leave for the user. Logico-mathematical languages are certainly not a gift of nature: they are created by a thinking human being. But there is no room reserved for the creator, no mark of the user is planned. The situation is completely different for natural languages in which "man makes his own style". They necessarily show the mark of the speaker even when he is masked by what Perelman (11) calls the universal subject, who is still a subject.

The consequence is that the constructions that each type of language can use, even if they are rarely exclusive to one type, are completely different. I shall call "models" the constructions used by logico-mathematical languages and "schematizations" those used by natural languages.

A model is a closed and universally valid construction. It is closed because if one of its elements is modified, it is not enriched but becomes another model; and it is universal because it depends neither on the context of use nor on its user. Indeed, it is important to note that it is because this property of being both free of context and free of subject exists that the classical trichotomy syntax, semantics and pragmatics can be used.

On the contrary, schematizations are both open and partially valid. For example, if we consider the schematization which represents a landscape, we realize that the author stops only when he wants to, and we will never come across a sentence like "and this landscape is nothing except through what precedes it". Even the word "schematization" clearly shows that we are confronted with a process and a result at the same time. And the limitation of its validity depends directly on the speaker's presence, on the "I", a theoretical concept not to be mistaken with any concrete subject, and which is nevertheless probably situated and located (6). It is understandable that the trichotomy mentioned earlier is inadequate here and that integrated pragmatics is more enlightening (10).

These differences, since they are significant, could lead one to oppose logic and language as two mutually exclusive worlds and between which a choise has to be made. Indeed, is it not completely different to accuse someone of being illogical, and of making a grammatical mistake? However, I don't think the situation can be so easily judged. First of all, as I said, logico-mathematical and natural languages do not have mutually exclusive uses: the most strictly proved theorem contains everyday expressions and the most surrealistic poem never gives all its words meanings completely different from everyday usage. Above all, because of their different functions, these two types of language do not have the same role in the elaboration of knowledge. Natural languages make it possible to reformulate knowledge in order to make sure it is coherent and fixed for a while. The two types of language are complementary tools.

1. REPRESENTATIONS

The term "representation" covers many realities. I will isolate two. First of all, what the subject imagines, which I will call "mental representations", and then what the subject represents through a system of signs, or simply what he presents. In this case one can speak of semiotic representations and in particular of "discursive representations". The distinction between the two sorts of representation is important, in so far as the observer has no direct access to the mental representations of subjects exterior to himself. They can only be inferred on the basis of clues given by the semiotic and - as below - discursive representations.

Having clarified this, we should ask ourselves what exactly the pedagogical art of teaching someone something is. At first glance, one can say that it is the act of substituting - by means which will be detailed later - one mental representation for another. A paradigmatical case is that of the missionary trying to substitute Christian faith for pagan beliefs by his preaching and personal example. We can go even further. The observation of Christian practices in non-Western civilizations shows that it is not merely a case of simple substitution. A fairly detailed analysis shows us that we are not observing merely the substitution of one representation for another, but the transformation of the representation which existed originally into one which is as near as possible to the intended one.

This is an essential aspect of any pedagogy. A teacher who knows nothing of a particular representation will not be able to transform it with any chance of success. In consequence it is very important for him to build a picture as near to reality as possible of the representations of his public.

This is not the place to discuss the large amount of research into the spontaneous representations of students and pupils. I will just remind you of an example in elementary dynamics (15) to underline two facts.

The first is that students always have some representation of the phenomena they are to study beforehand. These representations will not correspond to Newtonian physics, but they may well have a certain practical efficiency. Of course, they cannot be generalized, and some may even be mutually incompatible, but they are well adapted to the requirements of everyday living. This explains, at least partly, the second fact - that they are incredibly resistant to tranformation.

What is true for individual representations is also true for the collective representations which make up the sciences.

Transforming something - with or even without new elements - means restructuring what is present, and this cannot be done easily. For example, ever since the second half of the 19th Century, chemical scientists have known there are 63 elements. However, neither de Chancourtois'Spiral Theory (1862), nor Newland's law of octaves (1864) produced a structure which could explain known phenomena. They had to wait for Mendeleiev's (semiotic) representation (1869) for an organization which could be used adequately (9).

This example brings us to the question of how a representation can be transformed. Obviously, although Mendeleiev's table is not a discursive representation, it is accompanied by discourse. Of course, discourse does not really explain the emergence of the Russion chemical scientist's general idea - as does no other representation. The moment of discovery or invention cannot be explained by the use of signs. But it does not exclude the fact that - as Piaget showed (12,13) - the symbolic or semiotic (as he also calls it) function is necessary for reasoning to develop just as it is necessary for communication.

2. COMMUNICATION

It is not uncommon today to make a distinction between linguistic and communicative competence. The former normally refers to syntax and semantics while the latter - in a psychological or sociological context - refers to pragmatics. However, it must be quite clear that this distinction is connected with one particular conception of linguistics - the conception which considers (natural) language to be a system of signs, like a logico-mathematical language. On the contrary, if one considers that language is above all a social phenomenon and that - as I mentioned earlier - we can therefore speak of integrated pragmatics, then the two types of competence become one and the same : that of communicating mental representations by means of language.

In that case what does "communicating" mean? I shall define it as exchanging words in order to share some knowledge, which leads us - as can be seen - to the idea that every act of discourse is dialogic.

Of course objections can always be raised that situations of real dialogue seldom take place. The exchange of correspondence, letters from a reader to his magazine, are very specific forms of dialogue. Even when the two speakers are face to face all sorts of social constraints can lead to a pseudo-dialogue. All that is undeniable, and is not the point. Let us take the extreme case of someone sitting alone at his desk writing an article. It is obvious that he will not write the same article for a scientific publication, for an encyclopaedia, or a popular magazine. This means that he has different representations of his readers and writes according to them. He uses argumentation, and even if he doesn't actually use such turns of phrase as "you could always object that ...", he continually imagines the counter-arguments that could be raised. I realize there are times when the author is writing for himself only - in his personal diary, for instance. But even supposing he does not have the slightest intention of publishing it, he still uses a natural language, a language which contains "I" and "You". So a dialogue always exists, even if it is sometimes virtual.

This way of looking at discursive activity makes it very difficult to accept the traditional schema of communication in which sender A encodes information he possesses, transmits his message to receiver B who decodes it. This schema does make it possible to imagine that elements of information are present in the text - what Bailly called "dictum" (2), but not to imagine its "modus" or its argumentative and rhetorical components, to quote M.-J. Borel (4). I think it is better, therefore, to imagine the two speakers face to face. Each speaker takes a turn at speaking, builds up - in front of and for the other speaker - a discursive schematization of his representations, schematizations which the receiver in turn reconstructs. Instead of having the metaphor of the transmission of a message along a canal, we have that of the phenomenon of resonance - in the electrical sense of the word.

This explains the presence of the whole argumentative device in texts. The idea is to make it possible for the receiver B first of all to reconstruct what is proposed to him, and so we have titles and sub-titles, reminders, announcement of intentions, etc... But the receiver must also accept what he reconstructs, be able to forestall his own doubts and objections. All this generally refers to the "modus". Furthermore, in order to share something with someone, it must be understood, and here we have a fundamental problem.

Prieto (14) has emphasized that for a sign to become a signal it must be recognized as such by the receiver, in other words that he must discover an intention to communicate in the producer of the sign. In the case of language, the reader or listener must therefore recognize that he is in the presence of discourse. And that is only the minimal condition. He must also hypothesize that what is being said has a meaning, which is not always obvious. Even if we leave aside enigmas - in the strict sense of the word - whose study poses specific problems, it is not uncommon that the real meaning of an utterance does not coincide with its litteral meaning. This is obvious in the case of tautologies ("if he comes, he comes"), but such a phenomenon can occur anywhere. In this case, the listener-reader can understand and therefore reconstruct what is proposed to him only if he makes a real effort to try and discover the meaning.

As a result the problem is not so much one of interference on the canal but rather one of the isomorphism between the construction of a representation and its reconstruction. In fact, nothing suggests that it is ever complete. In the first place, A and B are not symmetrical, and there is only a very thin chance that their mental representations are identical. Such facts pose many delicate problems in teaching; they also mean that a text can be read in several different ways.

The double and necessary activity of the two speakers in a dialogue induces us to examine in greater detail the different powers of discourse.

3. THE DOUBLE-POWER OF DISCOURSE

I have just said that the activities of the speaker and the listener - or the writer and the reader - are not symmetrical. To construct a schematization from one's representations is not the same as to reconstruct one. Reading an oral or a written text has its own demands. "Schematically, the basic process consists in indentifying segments, formulating hypotheses, anticipating and checking" (p. 80), which do not correspond exactly to the operations needed to produce discourse. A and B should therefore be treated separately. In this paper, however, I shall take into account only general aspects which can be applied to both.

Globally, it is a question of knowing what the discourse activity of the participants contributes to the knowledge of the message. I think it has both an <u>organizing</u> and a <u>creative</u> power. In fact, these two aspects are closely linked. It may be possible to imagine an organizing activity, in which the creative component is limited to changing the order. Mendeleiev's table of the elements would be an example of this. On the other hand, it is quite impossible for something to be created without some need for organization. I mean by that, of course, the introduction of a completely new element to a body of knowledge and not simply the deduction of consequences implicitly contained in what is already known.

However, despite the fact that the creative and organizing powers are so closely linked, I shall try and look at them separately for the sake of clarity.

3.1. The Organizing Power

A schematization is elaborated through natural language, and, although - contrary to what happens in logico-mathematical languages - the users have great liberty of choice, they are bound by the language categories, by vocabulary at least. Colours have often been quoted as an example of how language contributes to organization - even as far as perception is concerned. But generally the need to elaborate one's knowledge so that it fits the mould of one's language gives it a certain type of organization. Anyone who has tried translating even once will know what I mean.

But that is not all. It may be possible to build systems in logic without negation, but it is definitely impossible to think without opposite pairs. "Night and day", "life and death", "hot and cold" carry as much meaning through the words used as through the things themselves. This is so true that their meaning is influenced to a greater extent by culture than by nature. Even the order to which these opposite pairs lead is not absolute. There is nothing to stop us introducing degrees of hot and cold, or talking about dawn and dusk, the first of which belongs to night and day, and the second to day and night. What I would like to emphasize is not that discourse imposes a certain order, but that there must be some order in the knowledge one wants to communicate.

It might even be possible that the "objectivity" of scientific knowledge is mainly the result of the need to communicate everything. In any case it is this putting into order that makes it possible to keep actual "living" at a distance from reflection. You only have to try and describe a dream you have had, or to give a discursive schematization of an activity as ordinary as walking, to realize the immense amount of putting into order you have to do.

The last point is a little more delicate, because it may depend more on the type of language being used than on a real necessity of reasoning. I am referring to the distinction between subject and predicate, an essential distinction for any Aristotelian logic. Of course, notions do not come to mind either as nouns or verbs (8). The idea, or notion of "going away" can become "departure" or "to leave". This indicates a basic organization of knownedge, even if, as Whorf states, certain languages have sentences with neither subject nor predicate (16). This problem is beyond my competence and should be studied in more detail since I can make myself perfectly understood in a restaurant simply by saying "coffee for one".

3.2. The Creative Power

Long before he is able to speak, a child can discriminate between objects, manipulate them, use them properly and we even know that Köhler's ravens distinguish between the first natural numbers when they are represented in a concrete fashion. It is therefore possible to hypothesize that at least some aspects of the extension of concepts do not depend on language. On the contrary - and this seems to me to be the greatest creative contribution of discursive activity - the intention of the concepts does depend on language. I can perceive colored objects, distinguish between red and blue ones, but if I want to give them a discursive representation, I must use abstract reasoning: I am going to "put together what goes together", use a relation of equivalence, I am going to speak of red and blue, and use a new abstract concept - that of color.

Finally, the demands of communication are such that all sorts of connections between objects, which would otherwise remain unsuspected, have to be explained. Explanations and justifications, hypotheses and consequences exist only if a schematization is built, and all that is created through discourse.

It is true that two major objections can be made to all this and to begin with Hamlet's "words, words, words". Words are not things and it can never be guaranteed that a discursive representation will be anything else than a representation of phantasms. As we well know, this danger does exist. However, we must remember two things. The first is that the powers of discourse do not exclude the necessity to have recourse to the powers of experience. We have every reason to use an active pedagogy, a pedagogy which gives the child and later the adolescent every opportunity to be face to face with objects and not words. We must also recognize, however, that we have all learnt more by reading and listening than by manipulation. How would we know anything of our past otherwise? We must not forget either that work on signs - and therefore on words - also makes experimentation possible, and this, not only at less cost, but also in cases where concrete experimentation is impossible. Counterfeiters are there to prove it.

Finally, one could object that the ability to speak does not necessarily infer the ability to understand. We all known the kind of pupil (often called "bright") who can recite their textbooks and formulae without error but whom we suspect of having understood nothing. The fact is undeniable, but, in my opinion, does not contradict what I have been trying to say. Reciting something does not satisfy all the conditions of a discursive representation. This is not an example of reconstruction, which is always the construction of meaning; it is merely a duplication of signs. I would never dream of saying that my tape-recorder, even if it had the whole of Kant's "Critique of pure reason" recorded on it, was a philosopher.

REFERENCES

1. Adam, J.M., et Legrand-Gelber, R., 1983, 'une société malade de ses lecteurs? in J'cause français, non? F. François, ed., Paris: Maspéro pp. 69-93.
2. Bally, Chs., 1942, Syntaxe de la modalité explicite, Genève: Droz.
3. Benveniste, E., 1966, Problèmes de linguistique générale, Paris: NRF.
4. Borel, M.-J., 1983, 'Sur l'activité de raisonnement', Travaux du Centre de Recherches sémiologiques, Neuchâtel, 44, a1-a26.
5. Cornulier, B., de, 1983, 'Logique spéciale ou pragmatique?', Intellectica, Paris, 7, 1-14.
6. Culioli, A., 1971, 'A propos d'opérations intervenant dans le traitement formel des langues naturelles', Mathematiques et sciences humaines, 34, 7-15.
7. Gilly, Y., 1983, 'Le texte et sa lecture' in Semen 1, Paris: Les Belles Lettres Annales de l'Université de Besançon, pp. 105-119.
8. Grize, J.B., 1983, 'Opérations et logique naturelle' in Borel M.J., Grize, J.B. et Miéville, D., Essai de logique naturelle, Berne, Francfort/M., New York, Peter Lang.
9. Histoire générale des sciences, 1961, R.Taton, éd., Paris, PUF, t. III, vol. 1.
10. Martin, R., 1983, Pour une logique du sens, Paris, PUF.
11. Perelman, C. et Olbrechts-Tyteca, L., 1970, Traité de l'argumentation. La nouvelle rhétorique, Bruxelles: ed. de l'Institut de sociologie.
12. Piaget, J., 1970, L'épistémologie génétique, Paris, PUF, Collection Que Sais-je?
13. Piaget, J., et Inhelder, B., 1966, La psychologie de l'enfant Paris, PUF, Collection Que Sais-je?
14. Prieto, L.J., 1975, Pertinence et pratique, Paris: Ed. de Minuit.
15. Viennot, L., 1979, Le raisonnement spontané en dynamique élementaire, Paris: Hermann.
16. Whorf, B.L., 1969, Linguistique et anthropologie, Paris, Denoël.

NON-VERBAL COMMUNICATION DEVICES : THEIR RELEVANCE, THEIR USE AND THE MENTAL PROCESSES INVOLVED

F. Lowenthal

Laboratoire NVCD

University of Mons

Non-verbal communication devices have been defined in previous papers (Lowenthal, 1978a, 1982). We want to show here, by means of examples, in what ways they are similar to and how they differ from any known verbal language; what makes them useful either for introducing new concepts in a teaching situation, or for creating a situation where the subject's behaviour can easily be observed and analyzed. Finally we will describe what mental processes are involved, according to us, when children are confronted with NVCDs. This can be illustrated by descriptions of experiments carried out with children (Lowenthal, 1979, 1984a). Two of these experiments are presented in this book (the first one concerns the use of NVCDs in a teaching situation and the second one concerns the use of an NVCD used to observe the behaviour of children solving a logical problem).

In order to do all this we will first recall the general definition of NVCDs, we will then explain why it is important to let children use communication systems which are associated with a strong logical structure that the user can easily perceive. It will then be possible to give examples where the usual verbal language - the most commonly used communication system - is <u>not</u> the most efficient: these examples will give us a list of criteria that would favour the development of communicative and cognitive abilities in special cases. These criteria will then be used to give a detailed definition of the concept NVCD. It will then be easy to explain how NVCDs can be used. We will finally formulate our hypothesis concerning the mental processes involved and let the experiments described further (LOWENTHAL and SAERENS, LOWENTHAL and HARMEGNIES) illustrate our theory.

1. A GENERAL DEFINITION OF NON-VERBAL COMMUNICATION DEVICES : FIRST APPROACH

A non-verbal communication device (NVCD) is essentially a tool (or a set of tools) which is furnished with technical constraints. These constraints make certain actions possible and others impossible: from these facts a logical structure emerges (Lowenthal, 1984).

We started developing the concept NVCD while working with handicapped children (emotionally disturbed and slightly mentally retarded); these children were apparently able to speak but did not master the complexity of the logic involved in the usual adult verbal language: teacher and pupils used different codes and did not understand one another, nor did they realise why this was the case. We noticed that the introduction of a semi concrete and highy structured material favoured the cognitice development of these children, at least in the field where the material was used: mathematics (Cordier, et al., 1975). We then wondered whether this could also be the case in other domains. Several clinical studies in classroom situations show that this is the case (Lowenthal and Severs, 1979; Lowenthal, 1978b, Lowenthal and Marcq, 1981, 1982).

2. THE IMPORTANCE OF COMMUNICATION SYSTEMS

A child needs to communicate with others. These "others" can be other children, or adults (parents, teachers, ...) who are called "educators" here. Such communication with others is essential to share or acquire new knowledge and skills: this remains true if we accept the idea that acquisition of new knowledge is based on imitation. In fact pure imitation is a (weak) form of communication, and more elaborated forms of communication can help the educators to focus the child's attention on what they want him to imitate and acquire as a new skill. A child also needs a communication device to communicate with himself (i.e. to think). It is vital for a child to learn sorting processes; it is also essential for him to be able to put labels on some of the facts he knows; these sorting and labeling processes will enable him to discover some regularities in his environment. One can expect that while doing this he will form his own new concepts. We think that this is exactly what BRUNER had in mind when he mentioned the use of several representation systems by a young immature human being (Bruner, 1966a, 1966b).

The most frequently used mode of communication by normal human beings is the usual verbal language. It is based on the audio-phonetic channel and is thus, generally, always available when two human beings are together. This verbal language can be used independently of what the interlocutors are actually doing (walking, holding a telephone in one hand a pen in another). This concerns the huge majority of human beings - but not deaf, or mute, people, nor patients whose speech has been impaired, ... Moreover this verbal language is not only very easy to use, but also very flexible: one can describe <u>all</u> possible actions, objects, feelings, ... with words.

One must note here that in fact the usual <u>verbal</u> language is not only based on words but also on others elements (intonation, context, ...). A single word such as "sure" can have different meanings depending on the intonation and the context. We give here three examples. At lunch time John asks Ann "Can you pass me the salt?"; she answers "Sure", which does <u>not</u> mean "<u>Yes</u> I can ..." but "I will do it". In the classroom the teacher looks at a pupil's answer and asks him "Are you sure?", the pupil usually interprets this question as meaning "You are wrong". In the bathroom, Ann asks John "Are you going to take a shower?"; he answers "Sure !!! There is no hot water", meaning "certainly not!".

These examples show that although verbal language is strongly associated with a logic, this association is not as easy to describe (and to decode) as it might seem. Many authors have described the logical structure which seems most appropriate for words only: ARISTOTLE was the first,logicians such as LEWIS CARROLL (1966) have given excellent detailed descriptions of useful axiomatic systems. But these axiomatic systems are not sufficient to describe all the intricacies of an existing verbal language: psychologists, linguists, sociologists, logicians and others: have been studying this problem for years ... and do not agree. They only agree on one fact we have already mentioned: every verbal language is associated with a logic. This implies an important question: how can children learn a language <u>before</u> they know the first elements of the logic associated with it and how can children learn the first elements of logic before they know the language which seems to be the only available description of this logic? This is obviously a vicious circle which all normal children eventually break, since they all eventually speak as adults do. How can they do this? We are confronted here with different interpretations of the data.

According to PIAGET (1936, 1947) children, though capable of verbal communication, are nevertheless unable to use purely formal reasoning before the age of 12 to 14: PIAGET believes that children start breaking the vicious circle by manipulating concrete objects without actually using a system of representation. BRUNER (1966a, 1966b) claims that all children operate with some kind of system of representation: manipulation of objects is one of them. The system adopted will be more or less complex and more or less powerful according to the level of the child's development his theory does not require the existence of several stages, hierachically organized, through which PIAGET claims that each child must pass with no possibility of hastening his development. SIEGEL (1978, 1982) has also questioned the standard Piagetian interpretation. According to her, children know more than they are able to express verbally: thus it is not correct to say that if children are not able to give the correct answer then they lack the associated concept; indeed, she proves that children can sometimes answer correctly, in a non-verbal way, where they fail if the question has to be answered verbally. RENARD (1985) considers language integration as a structuring activity and he insists upon the importance of structures as a basis for language integration. This is precisely what formal or mathematical logic is about: sorting activities first, followed by the discovery or the construction of relations which define structures in "the universe of discourse".

Logical prerequisites are thus needed, but it is obvious that they cannot be introduced via a verbal argument since these logical structures are part of the adult verbal language which is not yet mastered by the learner: the logical structures are thus, at the start, not yet integrated by the learner. The first elements of these logical structures should thus be introduced using devices which require as little verbal interactions as possible. Papert, although he claims to be a Piagetian, tried to use such a mode of communication: he created and used a computer language, LOGO, designed to give instructions to a real concrete object (a turtle). LOGO was invented in order to facilitate child-computer interactions so that children using LOGO would "invent" programming. PAPERT claims (Papert, 1980) that in the case of children who learned to use LOGO (almost) by themselves, concepts formation seems independent of the existence of the so called "Piagetian stages". This is also our point of view. Furthermore we think that the use of devices introduced in a nearly non-verbal way and based on techniques used in mathematics and logic, is essential; but we do not think that LOGO is satisfactory to start with: children must use too many words when working in LOGO (Harmegnies and Lowenthal, 1984).

What do we require from such devices? Firstly it should be possible to use them with as few verbal instructions as possible, but they should suggest in an obvious way some elements of the logical structure described by LEWIS CARROLL (propositional calculus). They should also be self contained and their meaning (if any) should be independent of any external component. Finally the number of relevant elements to be taken into consideration when creating a "message" should be as small as possible: this reduces the scope of such a device, but also reduces the load which is put ont the memory of the child using it. Since we began our research, we have used many devices inspired by these "rules of thumb"; we have used them with normal and handicapped children:

some devices appeared to be useful either to suggest logical structures to children an let them discover new concepts in a teaching situation, or to help us to place children in structured situations in which their behaviour could easily be observed and analysed. Some devices appeared to be useless: we thought they were perfect but the children did not agree! All this enables us to give a formal and complete definition of an NVCD. Before we do this, we want to end this discussion by stating how we think the vicious cercle mentioned previously can be broken by children. The experiments we did with NVCDs suggest that "the development of communication abilities and the development of cognitive abilities are interrelated; both depend upon the concrete manipulation of representations and upon the kind of logic these representations suggest". This last claim seems to place us somewhere between PIAGET's and BRUNER's theories (Lowenthal, 1980).

3. LIMITS AND FAILURES OF VERBAL LANGUAGES

One should notice here that we carefully avoid the word "language" when we mention concrete devices: a structured communication device is meant to be a concrete help for the user but restricts the range of the messages which can be expressed through it. Such a communication device is not and will never be a language: its main advantage is to clarify a complex situation. We give here examples where such a clarification cannot be achieved as easily through verbal communication only: these examples will suggest the sic criteria which serve as a formal definition of an NVCD.

The first example is easy: one does not teach someone how to swim by giving him a long lecture; one lets the learner swim and learn by his own body movements. Most adults can ride a bicycle, but not all of them can verbally describe what they do in that case. There is no need here for an artificial NVCD: the body movements are sufficent.

More interesting situations arise when we try to explain a "strange" game to a foreigner: how should an American explain the rules of baseball to a European, or an Englishmen, the rules of cricket to an American? Who is "in"? Who is "out"? What is "in"? What is the aim of the game? A European who reads several descriptions of "Baseball" might still be unable to understand this strange game: this is our case. It might appear useful to take us to an actual game, or to let us watch games on television. But this appears to be inadequate: there are so many irrelevant things going on (e.g. some people look pleased when another runs: is that part of the rules?). A better way to teach baseball to a European might be to show him diagrams, where the player's movements are shown by arrows. But this is static (although arrows indicating movements are better than frozen pictures). Baseball is action: let us use a dynamic representation. A computer can be programmed to accept only moves which respect the rules; sketches representing the active players and the ball, and nothing else appear on the screen. We could give our learner a special keyboard where each key is associated to a possible move, but we would not tell him in which order he should use them: the computer is programmed in such a way that keys which are pushed at a wrong moment provoke a negative reaction from the machine. We can furthermore programme the computer in such a way that it will keep track of all the attempts made by the learner, so that we can later analyze the learner's mental strategies when trying to make sense of what is shown on the screen (Harmegnies and Lowenthal, 1984). Let the subject start now: we can leave him alone and come back later. He will then explain to us all the rules he discovered (i.e. all the regularities he noticed); he will do so using a technical jargon he has created for the rules he has now learned to know, but he does not know our usual jargon. We also know the rules, and we know the usual technical terms so we can now tell him which label he has to pin down on which concept: he has now created the concept (using this computerized communication), this was not the case before and one cannot pin a label on a non existent concept. Each of us must first let things exist, then understand via his personal communication device that these things can be handled together and thus constitute a concept which immediately reeds a label.

A teacher might wish to describe a lattice and a polygon whose only vertices are lattice points. It is much easier to use grid paper and ask the children to draw the polygon; the instruction seems clear enough. However, anyone who has ever tried this exercise with children knows that problems arise (e.g. some try to use the centers of the square as vertices). A more experienced teacher will use a piece of paper where the lattice points are the only points to be seen: these are the only relevant points. By doing so this teacher has structured the situation, limited the possibilities for his pupils but carefully eliminated elements which, as far as the teacher is concerned, are irrelevant for the exercise on hand: this helps him to focuss the chidren's attention on what he wants to teach. Unfortunately, nothing prevents a child from drawing curves. How can a teacher make him understand that he cannot draw such things, without using a verbal negative statement? Let him use a geoboard with little nails to represent the lattice points (the only relevant ones) and rubber bands joining these nails to represent the only accepted lines: nothing has to be said the children cannot create a closed curve which is not a polygon with lattice points as vertices. Here there are many technical constraints which restrict the children's possible productions, but they do not feel that this is too constraining since geoboards can be used in a game situation. It then becomes easy to teach some basic notions of geometry very early or to observe how children build their polygons and discuss among themselves, in their own words, in order to find "the" best solution for a given problem (Lowenthal, 1984).

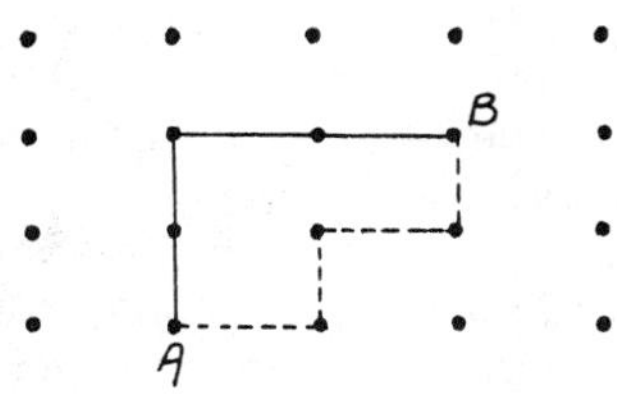

Figure 1.a. 2 paths of taxi-length 4.

Some people want to use taxi-cab geometry: this kind of geometry, imported from Manhattan, is done on grid-paper; the only accepted lines are horizontals and verticals, the only accepted points are intersections of accepted lines, the distance between two points is the length of the shortest path between these points, measured "as the taxi-cab goes" i.e. along the grid lines (see figure 1.a). A taxi-circle is a set of points at equal distance from a given point called center. On figure 1.b one sees a taxi-circle of radius 1. This circle is a set of points which are not attached to one another by any kind of straight or curved line. But all this is tiresome to explain verbally. We could simply use LEGO bricks. These bricks are designed to be put on a base board. The original idea is to build houses, but the bricks are so designed that one can only do one of 3 things: put one brick on top of another, next to another and in the same direction or next to another and perpendicular to it. This means that the only possible geometry with LEGO bricks is the taxi-cab geometry. This means that all the tiresome axioms about taxi-cab geometry are built-in in the bricks. It is thus reasonable to suggest that one should let children play with these bricks in order to let them discover, state and prove theorems concerning taxi-cab geometry (this geometry - more generally known as "absolute geometry" - is very useful in higher mathematics).

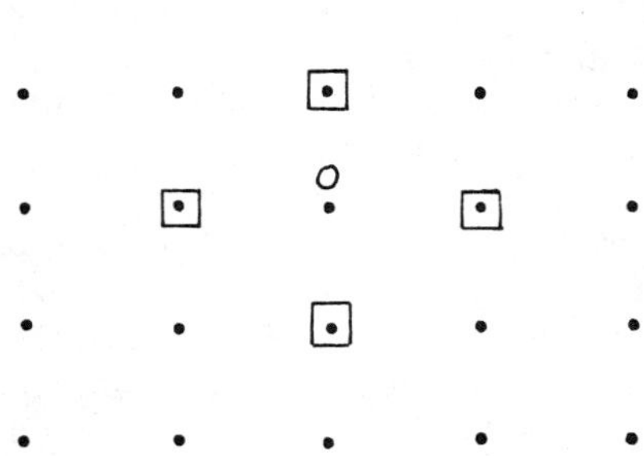

Figure 1.b. A taxi-circle of radius 1.

We can summarize all this noticing that although some games, or theorems, might seem very complex when they are formally described via the usual verbal language, they are in fact easy to understand, sometimes even easy to prove, when they are introduced via some non-verbal communication device with ad hoc built-in constraints. One must also note that reasonings and proofs are not less formal because they are presented without the help of technical verbal jargon: the main purpose for an educator is thus to prepare the child to look for an ad hoc communication device. This is not what is usually done: teachers keep referring to the verbal language as the only one which can be used for formal proofs, as the basic one to which the child should always refer (Laborde and Guillerault, 1984). Moreover, the use of NVCDs (e.g. LEGO bricks, geoboards, ...) enables the observer to see how a person deals with each step in the construction of a solution for a given problem; drawings can also sometimes make this possible, but the lack of technical constraints makes the comparison of observations more difficult: technical constraints provide a simple logical framework for the analysis. Observations can also be based on verbal interactions only, but as the subject generally does not know the technical jargon, this technique makes the analysis of the observations more difficult: this method should not be used unless the observer is interested in the knowledge the subject has of this technical jargon.

4. DETAILED DEFINITION OF THE NVCD CONCEPT

We have described several contexts in which a semi-concrete support, a non-verbal communication device, can favour the introduction of new concepts or the observation of subjects, or both simultaneously. Does this mean that whenever one uses a computer, a geoboard or LEGO bricks, one uses an NVCD? This is obviously not the case: a teacher who speaks about polygons and uses a geoboard (instead of a blackboard) to draw them, is certainly not using his geoboard as an NVCD.

Why should people use NVCDs? We already mentioned the problems met by young children: they do not use verbal language exactly as adults do and child-adult verbal interactions can thus often be misleading. Moreover, the verbal language used by adults is often ambiguous and its logical structures are not obvious: this fact makes it inappropriate for the sorting processes which the child needs while structuring his knowledge. To attempt to solve this problem we chose to introduce semi-concrete supports for the child's thought. Some were succesful and others appeared useless: the six criteria we are going to list now were not invented by a pure theoretician; they are the criteria common to all the successful devices we tried to use with children, one of them is always missing in the devices which have not been accepted by the children! (Lowenthal, 1980, 1982).

A device can be used as an NVCD provided one can thus confront the subject with logical problems in a nearly non-verbal way (i.e. with nearly NO verbal instructions) in order to avoid any conflict with a developing verbal language. This system should, by construction, avoid ambiguïties: the symbols used for propositional calculus in logic form such a system. Unluckily this system is too cumbersome and not suitable for children: two many non-obvious rules have to be respected to write a well formed formula. A good system should be simple and easy to handle. A single cube would be such a system, but it would be a very poor system: one cannot convey many different messages by means of a single cube. The simplicity of the device should not overrestrict the quantity of messages which can be formulated: the device must thus be flexible enough to let the people who use it have a high degree of freedom. This implies that an NVCD cannot be simultaneously totally non ambiguous, totally simple and totally flexible: one must try to create NVCDs which meet as far as possible these three requirements simultaneously. We only chose devices which could be used in the framework of games: this lets the children enjoy the session, feel secure and thus relaxed. Discovering new facts, putting old knowledge together to create further knowledge, create notation systems and start a long march towards abstraction and communication at a perpetually higher level of abstraction, all this is possible at the age of 6 but it should be a joy to do it, not a nightmare. Last but not least, in order to be an NVCD, a device must be full of built-in contraints which suggest a logic: it will thus be the system itself, and not the teacher who will "tell" the child: "You are wrong". Clinical data show that such negative reinforcements are more acceptable for children, especially because of the presence of an adult who will formulate positive reinforcements or why-questions which do not necessarily imply that there is a mistake: sometimes we might simply wonder whether the subject is able to verbalise the reasoning we have seen him using.

5. HOW SCHOULD ONE USE NVCDs: SOME EXAMPLES

The general rule is easy: first, let the children play freely with the device, they will learn to "feel" the role of the technical constraints; then challenge the child with a problem which is "just a little more difficult" than what he presently knows, but which is not "out of reach" so that the child has to create something new and can do it; eventually repeat this second step with gradually more difficult problems, so that the child can progress at his own pace.

A lot of things that are available in a regular classroom or playground can be used as NVCDs. New ones can be created. It is impossible to make an exhaustive list! We simply intend to mention here some devices whose use as an NVCD has been described in the literature; we will recall very briefly how they were used and state whether it was in a teaching situation, or to observe subjects involved in a problem-solving activity. We will add detailed references in order to enable the reader to create his own "user's manual" (if he so desires!).

the SOROBAN, or abacus, used by the Japanese (and the Chinese in a slightly different version) is probably one of the oldest devices which can be used as an NVCD to teach the basic facts of arithmetics. WEINZWEIG (1980) described a very simplified version: the bead frame. "This consists of 25 beads on a wire, the first 10 are red, the second 10 yellow and the remaining 5 red. The first 5 of each colour are one shade and the second 5 another ... Initially, all the beads are to the right and the child shifts them to the left as he counts". In fact in this context, the child only has to know the numbers 1, 2, 5 and 10 to start operating: this is exactly what happens with the SOROBAN. In the same paper WEINZWEIG describes other contexts favouring the introduction of a notation system where data are distinguished from results: the data are written in circles and the results in squares; the relative position of circles (empty or not) and of squares (empty or not) clearly indicates whether the question concerns addition or substraction.

WEINZWEIG also used another device, (not published). It consists of the frame shown on figure 2. He asked children: "Place the numbers 1, 2, 3, 4, 5, 6, 7 in the circles so that you have the same total on each of the three lines". Many different solutions can be found and the methods to find them are numerous; children can use trial and error or, for example, note that 2 numbers "are used twice" because of the intersections, that the sum of (1 + 2 + 3 + 4 + 5 + 6 + 7) is an even number and that the two vertical lines having equal totals, one can only use an even number for the central circle,... We feel that this frame could easily be generalised to other configurations where lines represent the technical constraints. We also think that a better approach would be the following: i) give the children the possibility of placing numbers on the frame and then to compute sums "along the lines", ii) ask them to compare the sums, iii) ask them to change the position of the numbers and

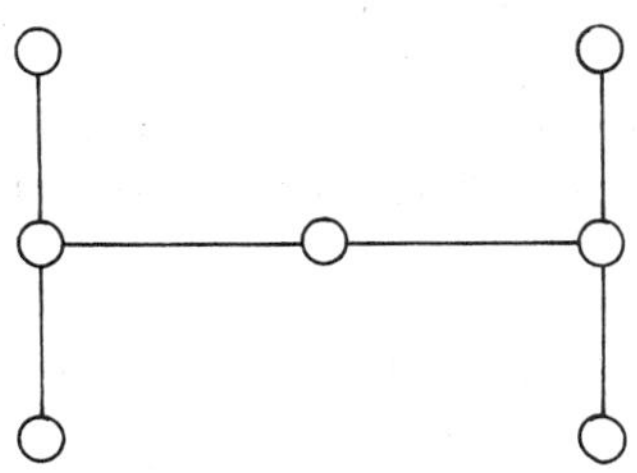

Figure 2. The big H.

compare the new situation with the previous one, iv) then ask the question used by WEINZWEIG.

CUISENAIRE rods can also be used as an NVCD. This material consists of "sticks built up from one to ten elementary cubes. CUISENAIRE has attached great value to binding the numbers to colours" (Freudenthal, 1973). The technical constraints here are restricted to the fact that one can replace a green stick (3 units of length) by three white ones (1 unit) or by one white one concatenated with a red one (2 units). In fact we think that the association "number-colour" is not as important as the association "length-colour": number is a by-product. These sticks can be used in many geometrical exercises not related with number concepts, but concerning only the structure of space. The sticks are mostly used in a teaching situation but could be used (as all NVCDs) to observe children involved in a problem solving activitity.

G. PAPY (1969) created a device he called the "Minicomputer" (see figure 3). This tool was devised to teach the basic arithmetical facts and is inspired by CUISENAIRE rods. "It consists of one or more 'boards', each board subdivided into four squares, and a set of 'checkers'. When one or more minicomputer board is displayed, the position that each board holds relative to the other boards corresponds to place value. The values of the four squares are: white is 1 (10, 100, ...); red is 2 (20, 200, ...); purple is 4 (40, 400, ...); and brown is 8 (80, 800, ...) ... A number represented by a configuration of checkers on the minicomputer is the sum of the values of all checkers on the boards" (F. Papy, 1977). The technical constraints reside in the rules used to handle checkers. Two checkers on a white square are equivalent to one checker on the red square (on the same board) and one checker on the brown square plus one checker red square are equivalent to one checker on the white on the square, on the next board to the left. "Although the minicomputer can be seen as a tool for calculation and as a device to help students learn routine methods for calculations, its more exciting use is as a vehicle for posing interesting problems which challenge a child's intellectual curiosity about numbers and for presenting situations which dually encourage strategic thinking and reinforce numerical skills" (F. Papy et al.

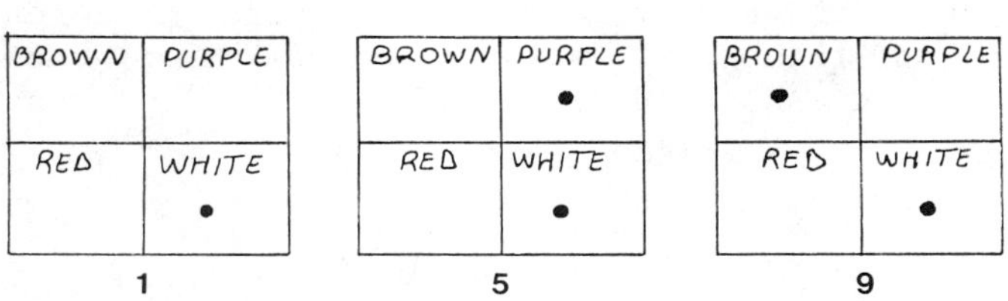

Figure 3. Minicomputer boards showing 159.

1979). This device is very often used in teaching situations but it has also been used by F. PAPY to observe children including "slow learners".

We have already mentioned that LEGO bricks and geoboards can be used as NVCDs. We used them to observe children and to introduce new concepts simultaneously. Further details can be found in another paper (Lowenthal, 1984).

DIENES created a set of blocks, "the Attribute blocks" which were meant to introduce basic logical notions such as "and", "or", ... (Dienes and Golding, 1970). The set of Attribute blocks (A-blocks) consists of plastic blocks; each block can be defined by 4 data: shape, colour, size, thickness. There are 4 possible shapes (rectangle, square, circle, triangle), 3 possible colours (red, blue, yellow), 2 sizes (big, small) and 2 thicknesses (thick, thin). In order to have all possible combinations one needs 48 blocks. This makes sorting and other logical exercises possible: one can introduce difficult exercises such as "make a sequence of blocks such that 2 adjacent blocks differ by exactly one criterion" or "cover exactly one big block by smaller ones" (Lowenthal, 1985). This device has been used either in teaching situations or to observe children.. In this book, LOWENTHAL and SAERENS describe how they used it in both situations simultaneously.

COHORS-FRESENBORG (1978) defined Dynamical Mazes. This device consists of bricks which can be placed on a baseboard. These bricks constitute the basic elements needed to create a railway network: straight rails, curves, by-passes and switches. This material has built-in constraints which purposely restrict the number and the kind of combinations a child can make, with the pieces: a "train" can only go through a "rail" in one direction, each piece is a "one way" element. As for the previous devices, the teacher does not need to tell the child that there are restrictions, the built-in constraints automatically impose them. A "Register Machine" based on similar principles, but implemented on a computer was also described by COHORS-FRESENBORG (1978), COHORS-FRESENBORG first wanted to use these tools to teach arithmetic and notions concerning finite automata. Later he became interested in children's programming attitudes. His basic method was to confront the children with a problem and to ask them to find the solution. We used the same material with younger children and chose to confront them with a solution and ask them what the problem was (LOWENTHAL and MARCQ, 1982; LOWENTHAL, 1984a). In fact COHORS-FRESENBORG chose to study synthesis behaviour while we studied analysis behaviour.

SAERENS (1985) described how to use a peg board to introduce the first notions of a programming language without a computer. We used this to introduce recursion and a pseudo-LOGO which is completely non-verbal in the first grade. Results concerning an aphasic child are described in the paper by LOWENTHAL and SAERENS included in this book. This device consists of a white plastic board furnished with holes. In these holes one can put coloured plastic nails. The shapes of the needs of the nails can be a square or a quarter of a circle (we wall these "triangles"). Using these coloured nails, children can make a mosaïque. The triangles serve as name for a sequence of squares and a sequence of triangles is a programme (Lowenthal and Saerens, this volume).

It is also possible to combine manipulations of an NVCD and observation by a computer. Detailed results concerning this aspect of the NVCD methodology are presented in the paper by LOWENTHAL and HARMEGNIES included in this book. This last kind of use of NVCDs is quite different from the ones mentioned above: it is essentially meant to observe in more details how children perceive the data of a problem. On the contrary the research done by COHORS-FRESENBORG, LOWENTHAL, MARCQ and SAERENS concern teaching and observation simultaneously.

Although we could mention many other NVCDs, we wish to conclude our long list here: this does by no means mean that there are no other valuable NVCDs. In fact there are many NVCDs, they have been used for ages but have not until now been identified, or grouped, as the same kind of tools, with the same characteristics and probably the same effect on cognitive development.

6. THE MENTAL PROCESSES INVOLVED

NVCDs seem to be fruitful in many domains. Why is this the case? We want to describe here a working hypothesis which might give an answer to this question. We want to illustrate by an example why we believe that our hypothesis is correct, although it cannot yet be considered as proved: we do not have statistical results concerning the teaching situations - further studies are being carried out at the moment but a clinical study is presented in the LOWENTHAL and SAERENS paper - on the other hand we have many statistical results concerning observations of children, some of them are presented in the LOWENTHAL end HARMEGNIES paper. All these results seem to support the hypothesis we are about to formulate.

Formulation of the Hypothesis

"Providing children with an NVCD introduces a structuring element in the perception of data. One might assume that the introduction of such a device into a child's universe serves as starter for a complex cognitive process".

i.e. Without the NVCD, the child would progress, but more slowly. The technical constraints of the device enable us to present him logical challenges and the structure suggested by the device enables the child first to apprehend the nature of the problem and then to organise a solution. In this way, an NVCD might favour the child's cognitive development and help him, to go from one so called "Piagetian stage" to another.

We assume that the complex process initiated by the introduction of an NVCD consists of 6 steps:

1) structuration of the perceptive field (in function of existing pre-concepts);
2) discovery of the relevant elements;
3) building of (local) relations between some of these elements, implicit formulation of the relevance of such relations and experimental verification;
4) building of global relations between all relevant elements, implicit formulation of the relevance of such relations and experimental verification;
5) verbal formulation, and proof, of a law concerning the theoretical functioning of the elements previously considered;
6) progression towards an abstract structuration of the perceptive field which enables the child to adapt his behaviour to new situations, in a similar context, but in the absence of any NVCD.

Illustration

By way of example, we shall describe the shifts in levels of relevance that we observed with 7-year olds who were discovering the concept of "finite automaton". As NVCD, we used the dynamical mazes described by COHORS-FRESENBORG.

Firstly, the children played with the bricks; they sorted the bricks when they discovered that there were different kinds of bricks: straight rails, curves to the left, curves to the right, crossings, switch points, ... The children tried to assemble these bricks and discovered that there were technical constraints. All these facts illustrate our first step: the structuration of the perceptive field which made possible the observation and perception of data.

We then asked the children to reproduce a network, whose sketch was shown to them, with their bricks. The children discovered that the constraints were important, that they imposed a given orientation for each piece: it seemed specially difficult to place crossings in the correct fashion. The children were thus led to the discovery of the relevant elements of each piece and had an opportunity to reinterpret the basic data.

When we asked our pupils to test their networks in order to interpret what they had done ("What is this network good for?") they looked at the list of outputs. Each network had one single entrance but several outputs and the "train" which ran through the network was forced to go to one of these outputs by the position of the switches; but while going through the network this "train" could modify the position of the switches in order to prepare another situation for the next train.

The children succeeded in making short term predictions ("Where will the next train go?") by looking at the regularities in the list of outputs. They had established local relations and made another step forward: from now on a relevant element was "a track", i.e. a path through the network from the entrance to one of the outputs. The children made hypotheses and tested them: they were now able to generate hypotheses.

All was well because we presented them easy, regular networks. Later we produced diagrams implying apparent irregularities at the level of the outputs: the problem could only be solved by making a list of all successive positions of the switches. This made it possible to notice for example that e.g. position 1 and 2 would never appear again, while positions 3, 4, 5, 6 would keep coming up, in that order, as if we were in a loop. At first, the children were puzzled, then 7 out of 20 succeeded in using this list of positions: they had now a global view of the network, they were able to make long term predictions concerning the outpost corresponding to the nth train. There was thus another shift in their level of reflexion and the children went beyond the obvious information presented.

Later these children were able to formulate the rules imposed by a given network verbally and to predict what was going to happen without actually using the NVCD. Then they stopped using the NVCD to give only correct verbal answers. They also formulated a proof and explained verbally whay "it works like this" and not otherwise.

At the moment nothing enables us to say now that these children could have initiated the process without the NVCD i.e. nothing enables us to say that they can transfer what they know to another domaine presented otherwise but nevertheless similar: the children have acquired the concept of finite automaton but they are not necessarily able to use this knowledge within the framework of a purely formal reasoning.

7. CONCLUSION

The examples we have given in this paper prove that many different materials have been developed which can be used as NVCDs as well either in teaching situations, or in order to observe children involved in problem solving activities. Most of these materials were created, sometimes a very long time ago, as pedagogical aids: the authors tried to create something to help pupils learn. The examples we chose for this paper mostly concern mathematics education, but devices having similar basic properties exist for the teaching of languages, geography,... All these devices can be used to observe children. When these devices were created, they all appeared to be a "new and essentially different" creation. In fact all of them are instances of the concept of NVCD. The notion of non-verbal communication device appears thus to be a general unifying concept.

An NVCD is certainly not a language, but a means to create a situation where sound structures communication can start, without ambiguity, and about a precise topic chosen by the creator of this precise NVCD.

The problems met by young children because of their insufficient knowledge of verbal adult language, suggested the use of devices which meet six requirements. The clinical experiments made with such NVCDs show that they can fruitfully be used for teaching and observation. This in turns suggests an hypothesis concerning the mental processes involved: this hypothesis is illustrated by the experiments described in the two following papers.

REFERENCES

Bruner, J.S., 1966a, Toward a theory of instruction, Harvard University Press, Cambridge (U.S.A.).

Bruner, J.S., et al., 1966b, Studies in cognitive growth, John Wiley, New York.

Cohors-Fresenborg, E., 1978, Learning problem solving by developing automata networks, Revue de Phonétique Appliquée, 46/47, 93-99.

Cohors-Fresenborg, E., and Griep, M., 1982, Registermachinen und Funktionen, Osnabrücker Schriften zur Mathematik, Universität Osnabrück.

Cordier, J., Lowenthal, F., and Heraux, C., 1975, Enseignement de la mathématique et exercices de verbalisation chez les enfants caractériels, Enfance, 1, 111-124.

Dienes, Z.P., and Golding, E.W., 1970, Les premiers pas en mathématique: logique et jeux logiques, OCDL, Paris.

Freudenthal, H., 1973, Mathematics as an educational task, Reidel, Dordrecht.

Harmegnies, B., and Lowenthal, F., 1984, Dispositifs de communication non-verbale et ordinateurs, Humankybernetik, 25(3), 115-124.

Laborde, C., and Guillerault, M., 1984, Learning to formulate in mathematics: imitation or creation?, in Language and

Mathematics - interactions between their developments , ed. Lowenthal F., Laboratoire N.V.C.D., University of Mons, 47-60.

Lewis Carroll, 1966, Logique sans peine, Herman, Paris.

Lowenthal, F., 1978a, Logic of natural language and games at primary school, Revue de Phonétique Appliquée, 46/47, 133-140.

Lowenthal, F., 1978b, Logic and language in game setting - for children ages 8 to 10, in: Proceedings of the second international conference for the psychology of mathematics education, eds. Cohors-Fresenborg, E. and Wachsmuth, I., Osnabrücker Schriften zur Mathematik, Reihe D., Mathematisch-didaktische Manuskripte Band 1, 214-225.

Lowenthal, F., 1979, Hypothetico-deductive reasoning at the age of 8 - clinical observations, in: Theory of Knowledge and Science Policy, eds. Callebaut, W., De Mey, M., Pinxten, R. and Vandamme, F., Communicatie en Cognitie, Gent, 149-159.

Lowenthal, F., 1980, Games, logic and cognitive development - a longitudinal study of classroom situations, Communication and Cognition, 13, 1, 43-63.

Lowenthal, F., 1982, Example of auxiliary formalisms used to help the development of children's logical thinking, in Language and Language Acquisition, eds. Lowenthal, F., Vandamme, F. and Cordier, J., Plenum Press, New York, 113-121.

Lowenthal, F., 1984a, Production langagière d'enfants manipulant un dispositif non-verbal de communication, Revue de Phonétique Appliquée, 69, 11-46

Lowenthal, F., 1984b, Development of communicative abilities and the acquisition of mathematical concepts in children, in Language and Mathematics Interactions between their developments, ed. Lowenthal, F., Laboratoire N.V.C.D., University of Mons, 19-60.

Lowenthal, F., 1985, Non-verbal communication devices in language acquisition, Revue de Phonétique Appliquée, 73/74/75.

Lowenthal, F., and Severs, R., 1979, Langage, jeu et activité mathématique - un essai à l'école primaire, Educational Studies in Mathematitics, 10, 245-262.

Lowenthal, F. and Marcq, J., 1981, Logic Auxiliary Formalism and Geometry by Telephone Call, in: Proceedings of the 5th conference of the international group for psychology of mathematics education, Vol. I, eds. Comiti C., and Vergnaud, G., Laboratoire IMAG, Grenoble, 265-270.

Lowenthal, F. and Marcq, J., 1982, How do Children discover strategies (at the age of 7)?, in: Proceedings of the 6th conference, Psychology of Mathematics Education, Ed. Vermandel, A., Universitaire Instelling Antwerpen, Antwerpen, 287-292.

Papert, S., 1980, Mindstorms, Basic Books, New York.

Papy, F., 1977, Math Playtherapy I, Cemrel, St. Louis.

Papy, F., Heidema, C. and Schweitzer, J., 1969, CSMP Mathematics for the upper grades, Part II, Teacher's guide, Cemrel, St. Louis.

Papy, G., 1969, Minicomputer, Educational Studies in Mathematics, 33(2), 3-45.

Piaget, J., 1936, La naissance de l'intelligence chez l'enfant, Delachaux and Niestle, Neuchätel.

Piaget, J., 1947, La représentation du monde chez l'enfant, Presses Universitaires de France, Paris.

Renard, R., 1985, Structuro-globalism and autonomy, Revue de Phonétique Appliquée, 73/74/75.

Saerens, J., 1985, Semantic components and syntactic sequences introduced by means of mutlicoloured plastic pegs, Communication and Cognition, Vol. 18(1), 24-28.

Siegel, L., 1978, The Relationship of Language and Thought in the Pre-operational child: A Reconsideration of Non-Verbal Alternatives to Piagetian Tasks, in Alternatives to Piaget - Critical Essays on the Theory, eds., Siegel, L. and Brainerd, C., Academic Press, New York, 43-67.

Siegel, L., 1982, The discrepancy between Cognitive and Linguistic Abilities in the Young Child, in Language and Language Acquisition, eds. Lowenthal, F., Vandamme, F. and Cordier J., Plenum Press, New York, 31-38.

Weinzweig, A.I., 1980, Concepts, Language and Context, Revue de Phonétique Appliquée, 55/56; 267-280.

COGNITIVE PEDAGOGY

OR A STRICT NOMINALISTIC APPROACH TO PEDAGOGY

F. Vandamme and R.U. Ghent

Seminaire Logika en Kennisleer
Blandijnberg 2
9000 Gent
Belgium

1. INTRODUCTION

How important in the educative process are language deviations and language disturbances? Are we dealing here with something marginal to the educative process or are we rather, immediately at the heart of the matter? In other words is the contemplation of language deviations and disturbances a phenomenon lying at the edge of the whole educative process? The answer to this question will determine how one handles language deviation and its treatment. If one considers it as a marginal phenomenon, then language deviation is a matter relevant only a) for a few specialists who as or with logopedists must help the deviating child when his deviation becomes very large or hindering, or b) it is the business of the general educator, when he has spare time or has realized his global objectives or as a breather, to hold himself busy with the language of his patients, his children.

Diametrically opposed to the situating of the language problematic as marginal stands the vision that symbolization in general and language symbolization in particular are central to each human action-process and central therefore to education of any kind (pedagogical, orthopedagogical or orthoagogical). In this perspective educative action can be meaningful only when one at the very least takes account of the symbol-environment of the actors (educator, educated) and via symbols adapted to this environment, acts and interacts. Deviation in the generating or recognizing of symbols, of what ever kind, will bring disturbance to the educative process, and vice versa. In this vision the commission of language deviations or disturbance (language is one of the most sophisticated of the social symbol systems) is central to every educative process. Symbol application and symbol analysis then are in this vision also the beginning and the end of every educational process. This vision can be termed cognitive pedagogy of orthopedagogy. But it is clear that here cognitive must be taken in a broad sense as embracing aspiration, will and purpose connation, together with planning and intellect in its broadest sense (see Vandamme, 1985).

If we now fellow Protagoras and thus lay every emphasis on the symbol as a socio-individual product originating from a) the interaction and coordination of the individual with his co-actors and from b) the exchange and coordination of the various subaction fields within the individual, then there are undoubtedly far-reaching epistemological, genetic and ontological consequences. We do not wish here to go into these aspects[1] .Rather shall we merely signal that in this perspective no phenomenon is relevant unless it is symbolized directly or indirectly by one or another actor in the process in one or another component (wherein the material component of the symbol can be more or less important). This is another way of formulating the "man is the measure of all things" principle.

If we now from within such strict nominalistic framework take a look at education, we can pose the traditional structural questions:

1. What is the basic situation: the situation at the commencing of the education process?
2. What is the target situation?
3. What are the means?
4. What are the efficiency or adequacy reguirements?

What answers to these questions proceed from within a strict nominalist framework? This is of especial interest to us. We want to find a formula which provides a start-to-finish answer to these questions.

2. THE STRICT NOMINALISTIC DESCRIPTION OF THE PEDAGOGICAL SITUATION: A REGISTER APPROACH

The education process is based on an interaction between educator and educated in a specific situation. It is clear that this process can occasion actions in a classroom which will differ from those on the football field or in the street. Much here will depend on the sort of actions and the roles performed by the actors in the process. A technical lesson during which everyone works with a piece of metal clearly differs structurally from a class in which the teacher sets out an explanation on the board to which all must attend.

It is also obvious that much attention is given in the pedagogical literature in general and the orthopedagogical in particular to the components: the actors (educator and educated) and the situation.

More and more authors point out that in the analyses of these components there is the danger of unjustified simplification and overgeneralization. In labelling especially lies a similar danger[2], which in traditional diagnostics comes to a head over deviating characteristics of the child f.i. Opposed to this, one can assume that each diagnosis must take account of the three basic components and their interrelationships. The danger however, is that one component will dominate the other. Van Schoorl (1981, p. 81) even indicates that labelling is often used as justification for a specific organization or institution (situation), and in this sense handicaps, for example for keeping the concerned institution in existence. To illustrate: "The school now legitimates its existence by referring in its name to the existence of a category of children. The children must then be labelled and after the distribution of the label the purpose of the school concerned stands fast by definition".

Van Schoorl ultimately criticizes this approach. He himself suggests, following Ter Horst, placing the adult, the educator central rather than the child.

"Ter Horst sees the beginning of educative questioning as 'actionshyness'. If people reflected on their interaction with the child, they would sometimes ask themselves: 'What must I do?' We're not talking here primarily about a problem of children, but a problem which confronts adults. It is mainly implicit, felt as an irritation or a concern when something arises which is not easy to deal with. It is more easily termed a problem in education if one becomes aware of it; and the choice made among various possible solutions will influence the direction of the child's development. Pedagogically seen, one views the problem as being explicitly one of choice for the adult, who from the standpoint of his responsibility looks at the developmental paths and the developmental chances of the child. Directly bound in here is the vision of the adult of the desired direction of that development.

This means that labelling in this perspective is not a characteristic of the child but rather the opposite: the problem of choice for the adult, the educator.

But one remarks, are the adult's problems of choice in relation to the development and chances of the child independent of the child in its situation. Naturally not, which is why we see that each attempt to approach education in general or the orthopedagogical problematic in particular by privileging one of the three important components is not only arbitrary but also very dangerous in that so many important components and information central to the hypothesis and its answer are eclipsed.

The conclusion lying to hand is that one must view each and every one of the above-mentioned components of education, viz. all the actors (educated and educator) and the situation. But it is clear that with this much and little is being said. Much in the sense that a foundation is laid, but little because no indication at all is given of how the foundation should be filled in. From within strict nominalism one can attempt to realize this by applying a register technique, as follows;

1. First and foremost one tries to determine the various relevant processes in which the chief actors are involved (eating, playing, going to the toilet, sleeping, singing, following lessons, running etc.).
2. One tries to make explicit the verbal and non-verbal symbols which are used in each specific process (specified in 1). Any such worked out system of symbols is called a register (or model).

With each such process we can relate several differing registers. First we must make a distinction between concrete and abstract registers. A concrete register describes interrelations among symbols as they are actualized in a certain actual action-process. An abstract register on the other hand is constructed from a series of analogous concrete registers and it models a structured collection of potential symbolic interrelations.

Secondly, we can differentiate individual, group or cultural registers. An individual register describes symbolic interrelations, in other words a model of a certain action process from the perspective of a certain actor. It is obvious that according to their role in this process, specific accents, relations and specializations will appear in the individual registers.

A group of cultural register reflects the relations of the symbols in a certain process as experienced by a certain group or culture. Naturally a question about concrete technique arises, viz. how these various registers can be worked out. Furthermore, a whole series of methodological and epistemological questions must be raised. But we don't intend to discuss them here, except insofar as it is clear that each register which is built can always be questioned and remains open to adaptation. Adaptation can be needed because either the basic process have not been rightly differentiated, or the registers have not been satisfactorily worked out (insufficient data), or certain interrelations are inaccurately reflected, or finally, the process itself had evolved.

The value of a register construction must always be established as a function of its adequacy requirements. These will naturally be a function of the purpose posed with their construction. In a pedagogical perspective these adequacy requirements should naturally be pedagogical, for instance allow for intervention in the actions of educated or educator.

Taking all this into account, one can attempt to build the relevant registers in a concrete situation, registers which give us a reflection of and insight into the symbolic vision which the educated in this situation (including interaction) has of the eudcator, and eventually of others; or the vision the educator has of the educated. At the same time we can attempt to fix group and cultural visions in relation to their relevant processes through the agreed-upon group and cultural registers. Having realized this, we can naturally place them in confrontation, locating discrepancies, oppositions, complementarities; predict potential conflicts, declare real conflicts and introduce the strategies for avoiding them etc.

In practice the register approach implies that three important types of series of registers exist in the pedagogical situation. The first type is the register series of the educated. The number of registers in this series depends on the activities in which educated are involved. Their registers point out the different actors, but from the perspective of the educated. Thus the educator has a place there. We can also find ourselves with super registers which cover various registers, which does not imply there are no contraditctions among registers and among super regisers.

The second type of register series is that of the educator. This series again contains registers in which the educator is involved, but each register is constructed from his perspective, although naturally place is given to each other actor in his action field.

The third type is the register series of the culture group, in which the pedagogical situation of the concerned actions is embedded.

It is not thereby impossible, rather it is apparent, that a certain symbol which has a specific meaning for the one group as sender, will have a different meaning for the other as receiver.

3. THE EDUCATIVE PURPOSE IN A REGISTER PERSPECTIVE

Ideally we should be able to model the educative purpose through a fourth series of registers, in which the other register types are mingled in harmony. This means thus that educative action presupposes an operation on the educated, the educator, the situation and the culture. This last component, namely that the educational transformations can imply also an transformation on the culture itself, is seen repeatedly in many publications. It is clear that such action is one of the most difficult and uncertain in which to obtain results.

In practice one limits himself to action on the educator, the educated and the immediate environment in which both interact. The intention hereby is that the registers of those involved and therefore also their action will be ameliorated. We want to stress that by changing the registers also the action is changed. For the actions these are naturally (as with every creature of a certain minimal development) function of the models (registers) constructed by the actors. In other words the educator, through his specific interaction, aims to bring the development of the models, the registers of the educated, to a higher level and thus his action. But one can remark that it is easy to describe the purpose of the educator as being the achieving of a higher level of development of the models (the register containing verbal and non-verbal symbols) of the educated,. Rather the question is to determine this higher level. Which adequacy requirements can be formulated, which efficiency criteria posed for the education strategies in this perspective.

4. EFFICIENCY MEASURES OR ADEQUACY CRITERIA FOR REGISTER TRANSFORMATIONS:

We can approach this problematic from two angles.

The more subjective one poses as central the internal situation (desires, feelings) of the involved persons. The other method poses rather the functioning in his environment of the educated, independent of these internal situations, as being central. Naturally one can attempt to combine both. Metaphorically one should term the first approach the subjective or internal approach and the second the objective or external approach [3].

4.1. External or Functional Approach

One could say that progress is being made by the educated in their education process when:

1. an increased complexity is to be observed in the actions of the educated, their products and symbols used. Such increase in complexity can be established and even measured by determining the increased complexity of the correspondent registers and subregisters. In this vision increased complexity is thus identified with progress. It is clear that this is a rough approach. But on the other hand, from experience, we know that progress usually implies more complex structuring, although not every increase in complexity, signifies progress (cf. Morin, 1977).

2. Decreasing involvement of the educator , keeping the complexity constant on even increasing it. Another important possibly

complementary, criterium which can be introduced is decreasing dependence, and this in opposition to the normal dependence-strengthening by the educator. As by many authors (Illich, among many others) one can emphasize that institutes, systems, tend towards selfpreservation. One of the important strategies for this, mostly unconsciously brought into operation, is that he who helps, makes the one he helps so dependent on help that he can no longer go without it. Therefore it is very important to fight this. In this perspective one can call successful any educational treatment to the extent that the educated are helped in achieving a higher level of development, but thereby learn also to achieve a certain rise in independence. Naturally it is not denied with this that each person is a social creature and because of this dependent on the other, the group. What is intended with this criterium is not that the educated becomes detached from the other, the group, but that he interacts precisely with the other, the group, while becoming detached and independent from the educator - in other words increasing independence from the educator is demanded. An increasing independence which may not signify however, and certainly not in long range terms, a decreasing level of interaction with the other, or a decreasing complexity of the corresponding registers.

3. An increasing adaptation to the environment of the group. The interaction with the other can enlarge the level of the educated (achieve greater complexity), but the danger exists that thereby the inadaptation becomes greater too. But at midrange the mutual adaptation must be satisfactory. We can express this as follows. The development of the registers of the educated and the group must move in the direction of increasing congruency and/or harmony, at least partially, for a number of domains. But one has also in this to take into account that the educated has to retain his uniqueness his personality and yet to adapt so as not to become isolated.

This problem which is present in the relations between individual and group, is also posed analogously between cultures. In the interaction between cultures, one will threaten to dominate the other, to absorb it. Also from an ecological perspective a similar absorption is dangerous.

A culture may and must adapt itself to surrounding cultures, to the extent that it can continue its own development and retain sufficient independence from the other cultures so that it can adapt some of the givens from them for the achieving of its own goals. The great danger in this process is naturally that the culture loses its own personality, its own goals in this interaction, and simply becomes absorbed by the other culture.

We could pose something analogous about the relation between the individual and the group. When an individual's adaptation to the group brings with it an increasing intellectual and actional realization-potential, this is positive. When, however, the adaptation paralyzes the individual, possibly demoralizes him, then things are going wrong. Higher increase must be seen broadly. It must encompass the whole field of individual (or in the case of cultures, social) expressions and actions and may in no way be limited to the economic. Yet the economic must not be underestimated here. It is an important but a dangerous aspect of the human individual and social situation. It is important because there is an economic aspect connected to every action (be it sometimes secondary or even simply epiphenomenal). It is dangerous because it degenerates easily into becoming a goal in itself, subordinating all the rest.

We want also to emphasize that we interpret increasing self-development seen in the frame of strict nominalism and thus in the light of the rejecting of essence in terms of social functioning and realisation.

4.2. Internal Approach

1. We get a completely different angle on the efficiency of the education process if we take the standpoint that through this process the educated must find himself in the state of being better able to realize his goals, his desires. In this case the educational process can be evaluated in function of a) the means which are offered to the educated, b) the use of them which he is taught, and c) the competence he has to transform his goals so that he becomes able to realize his goals.

2. We could also use as criterium a lessening in the experience of discomfort or possibly (something fundamentally different) a rise in the enjoyment of pleasure of the educated. Possibly this can be coupled (as with 1)) to the realizing of socially accepted action patterns. We obtain then as a measure of efficiency the fulfilling of duties, roles with less discomfort (or more pleasure).

In the register model this will be established by a shifting (or possibly banning) of negative or positive (according to the case) evaluation symbols in certain registers or subregisters. So for instance certain actions (during reading, working) which are symbolized as negative in certain registers, in opposition to watching television which is evaluated as very positive, become instead evaluated neutrally or positively while the television-watching becomes neutrally or negatively evaluated...

5. ABOUT EDUCATIONAL MEANS

In a Protagorian/Piagetian viewpoint, education requires social/ individual interaction. In this way of thinking it is emphasized that in this interaction the instruments and materials (and from there their symbolizations) provided by the culture, the group, are primordial. And mustn't the properties of these materials, instruments, be discovered by the educated! Given the general importance of analogical reasoning, it can be assumed that a) these established properties and b) the successful strategies for finding them as well as the social appreciations and reactions thereto will streamline the model building and thus the forming of the individual in his various aspects (cognitive, emotive etc.).

From within a similar vision Papert underlines the importance of instruments, in his compelling book, Jaillissement de l'Esprit:

"J'accorde plus de poids que lui (Piaget) au rôle que jouent les matériaux offerts par telle ou telle culture dans la détermination de cet ordre. Notre culture occidentale, par example, est très riche en matériaux utiles à l'enfant pour mettre en place certaines parties de sa pensée logique et numérique. Les enfants apprennent que le résultat d'un dénombrement quelconque est indépendant de l'ordre et de la disposition des parties: Ils étendent cette notion de 'conservation' à leur réflection sur les propriétés de liquides quand on les transvase d'un récipient dans l'autre, et des solides quand ils changent de forme. Les enfants élaborent ces structures de pensée de manière préconsciente et 'spontanée', c'est-à-dire qu'elles leur soient volontairement inculquées. D'autre structures de la connaissance, comme l'agilité d'esprit que réclament les permutations et les combinaisons, par example, ne s'acquièrent que plus lentement, ou même ne se développent pas spontanément, et doivent être abordées grâce à un enseignement classique. Le présent ouvrage, dans son ensemble, tend à démontrer que la différence observée dans le développement de ces structures est imputable, dans bien de cas important, à la relative pauvreté de notre culture en matériaux qui permettraient de bâtir ces structures intellectuelles en apparence 'plus avancées' ".

Having underlined the importance of the instrument, Papert introduces the computer to the educated as the ultimate instrument. We shall not go further into this problemata. We wish simply to underline the importance of instruments (among others, toys) for the educated and the importance of the analogical operations or analogical logic gained through experience with this instrument.

NOTES

[1] . More about this in:
Vandamme F., et alii, 1982, Registers; kennisrepresentatie verwerving en verwerking nr. 1 werkgroep registers C & C.
Vandamme F. en Vandenbrande R., 1982, Waarheid en Kunst.
Vervenne D., et alii, 1982, Een duiding van het striktnominalisme, nr. 2.
Vandamme F., 1981, Language and cultural adaptation Unesco, Paris.
Vandamme F., 1983, Strictnominalisme, pragmatics and education.

[2] . Wens called our attention in this connection to, among others the reports of Rink (1981), Schoorl (1981) and Adriaans (1980).

[3] . We wish to move carefully in the use of terms as external/internal and objective/subjective because they suggest an unbreakable duality which to us is misplaced.

REFERENCES

Dupréel, E., 1948, Les sofistes. Editions du Friffon, Bruxelles.

Frazenburg: Ein Strüber - The learning of algorithmic concept by action. A study with deaf-mutes Revue des phonétiques appliquées. Vol. 55-56 pp. 249-253.

Joos R., Vandamme, F., 1983, Omtrent de rol van de "vraag" in het onderwijs of de vraag als hefboom tegen leermoeheid, in registers, pragmatiek en schoolmoeheid nr. 3.

Knorr, K., The manufacture of knowledge.

Löwenthal, 1980, Formalisme auxiliaire: pourqoi faire? Comment l'utiliser? Revue des phonétiques appliquées. Vol. 55-56, pp. 263-266.

Morin, F., 1977, La méthode, ed. du Seuil, Paris.

Papers, S., 1980, Jaillissement de l'esprit, Flammarion.

Spoelders, M., Van Besien F., Lowenthal, F., Vandamme F., 1985, Language acquisition & learning ACCO Leuven.

Spoelders, M., Van Besien F., Lowenthal, F., Vandamme F., 1985, Discourse, ACCO Leuven.

Vandamme F., et alii, 1982, Registers: Kennisrepresentatie verwerving en verwerking, nr. 1. Werkgroep registers Communication & Cognition.

Vandamme F., Language and cultural adaptation, Unesco, Paris.

Vandamme F., et alii, 1983, Symbolisation, worldmaking and the formation of teachers, in registers, pragmatiek en schoolmoeheid nr. 2.

Vandamme F., 1983, Strictnominalism, pragmatics and education, in Pragmatiek en Onderwijs, nr. 11.

Vandamme F., 1985, Introduction Cognitive science. CC-AI

Vervenne D., et alii, 1982, Registers. Een duiding van het striktnominalisme, nr. 2 Communication & Cognition.

LANGUAGE, LEARNING AND TEACHING:

HELPING LEARNERS TO MAKE KNOWLEDGE THEIR OWN

Gordon Wells

Ontario Institute for Studies in Education

At the moment, in thousands of classrooms around the world, teachers are talking. They are teaching and their pupils are learning. But whether the learners are learning what the teachers are teaching and, still more, whether they are learning because they are being taught are questions that do not have self-evident answers - despite the assurance of some curriculum planners and educational administrators that careful scheduling of the input is the only way to ensure satisfactory output. But is it appropriate to talk in terms of input and output at all?

In societies with universal education that have become prosperous on the basis of manufacturing and mass production, metaphors from industry are ready to hand: 'plant', 'resources', 'unit cost', etc. In the United States I recently heard two educators talking about 'tooling up the curriculum' for the following year. Presumably the school was the factory in which the new precision-engineered curriculum was to be installed, with teachers to operate it and children, the raw material to be processed into acceptably educated members of society.

In discussions among teachers, too, it is nearly always the input - that is to say the teachers' perspective - that is adopted in discussions of the curriculum. The traditional conception of education as 'cultural reproduction' doubtless goes a long way towards explaining this, as do the organizational problems experienced by teachers in ensuring that thirty or more pupils are kept profitably employed. It is an easy step, therefore, to move from a very proper concern to discharge these responsibilities effectively to believing that well-prepared input will lead to effective learning. The reality, of course - as every teacher knows -

is much less simple. But the majority still cling to the 'transmission' model.

Despite their basis in tradition and their amenability to assessment for the purposes of accountability, these input-output models are, I wish to argue, fundamentally misguided. Their basic flaw is that they leave out of account the contribution of the learner.Or rather, they treat the learner as a relatively passive recipient of teacher-structured material - as malleable substance to be shaped by the teacher's manipulation. To focus one's attention on the learner, however, is to arrive at a very different understanding of the relationship between input and output. Rather than clay in the potter's hands, pupils now appear more like potters themselves - apprentices working alongside a master craftsman, observing him put his knowledge to work and, as they try things out for themselves, attempting to make that knowledge their own.

This view of the relationship between learning and teaching is not new, of course. It is to be found at the heart of the writings of some of the greatest educational thinkers. But it has rarely informed the practice of the majority of teachers and, in the present climate of retrenchment characterized by the call for return to 'the basics', it is in danger of being almost totally ignored. In this paper, therefore, my intention is to examine more closely the relationship between input and the learner's response to it, with a view to clarifying the role of the teacher.

I shall start with a consideration of first language learning, since this is the most fundamental type of learning to require the contribution of other people. To a very large extent, of course, this learning takes place before a child goes to school and embarks on the more formal stage of his education. However, for a substantial proportion of children the learning of a second language is the first task that they have to engage in when they get to school, and all children still need to extend their linguistic resources, including the acquisition of written language. Language learning is thus a major focus of attention throughout the primary years.

As well as itself being a subject in the curriculum, language also constitutes the medium through which most of the other subjects of the curriculum are encountered. All too often ignored or treated as transparent, the quality of the linguistic interaction between teacher and pupil or between writer and reader has a major influence on the opportunities for learning that the classroom provides. As I hope to show, the way in which this is managed can make all the difference between an environment in which pupils are encouraged to make knowledge their own and one in which their attempts are impeded by having the handed-down knowledge of others thrust upon them.

The evidence on which I shall mainly base my argument comes from the study of 'Language at Home and at School' that we have been carrying out in Bristol over the last 15 years. During this time we have made a longitudinal study of a representative sample of 128 children, selected from a much larger random sample to give equal representation to both sexes, the full range of family background and to births in all four seasons of the year (cf. Wells, in press a for details). Two age-groups were involved : half the sample were aged 39 months at the time of the first observations, the other half 15 months. From this latter group exactly half were selected for continued study after they started school and our last assessment of them was made in 1982, when they were 10 years 3 months old.

Throughout, the study has been observational. During the preschool years, each child was recorderd ten times at three-monthly intervals, using a bugging device - a radio microphone worn by the child - that transmitted to a radio receiver, which was preprogrammed to switch on and off at approximately 20 minute intervals, recording 24 samples of 90 seconds duration over the course of a complete day between 9 a.m. and 6 p.m. As no observer was present, we are confident that we obtained representative samples of each child's normal experience of linguistic interaction within his family circle. Contextual information was obtained in the evening, when the recording was played back to the parents and they were asked to recall the relevant moments of the day in as much detail as possible.

A similar technique was employed when the children went to school. Each of the 32 children was recorded again, first before he or she started school and then three further times during the first two years at school. For these recordings, the sampling period was increased to five minutes, and nine samples were recorded between 9 a.m. and 12 noon. In the classroom it was found necessary to have an observer to take notes and, this being so, we decided to include a mechanical observer in the form of a video camera. Again however, the aim was to observe spontaneous classroom interaction and every effort was made to minimize the effect of the observer's presence. Here too, we believe that we were reasonably successful (cf. Wells, in press). Finally, we assessed the children more formally on entry to school and again after two and five years of schooling, using tasks and assessment schedules completed by the teachers. We also interviewed the parents and the class and/or head teachers at the same points and at 10 years we also interviewed the children.

Initially our concern in this longitudinal study was to chart the course of language development during the pre-school years and to look for environmental factors associated with differences between children in their rate or route of development. Latterly, however, we have been more concerned to identify the influences on educational attainment, particularly in so far as these have a linguistic basis. In the remainder of this paper, I wish to consider some of the more important of the findings of the study as they bear on the relationship between learning and teaching. Other aspects of the research have been reported elsewhere (Wells, 1980, 1981 a and B, in press a and b).

THE ROLE OF THE INPUT IN FIRST LANGUAGE LEARNING

For the last thirty years or so there has been quite a vigorous debate about the role of the input. Behaviourists (e.g. Skinner, 1957) have stressed the formative influence of the input, arguing that language learning is essentially a reproduction of the model of the target language provided by mature speakers, brought about by imitation on the part of the learner and by shaping of his linguistic behaviour by those on his environment. Some of the recent intensive studies of 'motherese' , as the special register adopted when talking to language learners has been called, have tried to give greater precision to the description of the teaching techniques that parents are supposed to use (e.g. Moerk, 1983). Nativists, on the other hand, have tended to belittle the role of the input. Chomsky(1965; 1976), for example, has argued that, equipped with a language faculty - a language acquisition device - all the child needs is to have the learning process triggered by exposure to experience of language in use. Several contributors to two recent edited collections (Deutsch, 1981; Wanner and Gleitman, 1982) have taken a similar line.

But are the two points of view totally incompatible - as their proponents seem to believe? As with most instances of the nature-nurture controversy, isn't the truth more likely to lie in some interaction between the child's innate predisposition to learn and the contribution made by his environment? This is certainly the interpretation I would put upon the findings of our longitudinal study.

First, we found that, despite quite wide variation in the sort of conversation that they experienced, all the children went through essentially the same sequence of development. This was true even at a quite fine level of detail. It did not appear to matter whether the parents devoted much time and attention to their children, talking and playing with them and sharing their interests, or whether they interacted with them relatively little, rarely giving them their sustained attention. In either case, the children still followed the same course of development. This strongly suggests that the input provided by adults has little effect on the sequence of learning. However, the finding with respect to rate of learning was very different. From this point of view, the effect of the input was quite considerable, with significant positive correlations between both quantitative and qualitative characteristics of the input (Wells, in press, a).

At first sight, these findings may seem to present us with a paradox: that input both is and is not of importance for language learning. However, there is only a paradox if we think that the learning of a first language involves the transmission to the learner of an adult model, with the sequence of learning determined by the selection, timing and reinforcement of the transmission, which is under the adults' control. It is true that, typically, when we talk about language learning in an educational context, we tend to think of it in terms of teaching: we think of vocabulary, structures, idioms, and so on, that need to be taught if they are to be learned. And when we talk about children's learning, what we typically mean is that they show evidence of having learned that which they have been taught. However, as I suggested at the beginning of this paper, I believe that this teacher-centred view of language learning is entirely inappropriate. Children are active constructors of their linguistic resources, not passive recipients of the language of others.

If we go right back to the beginning of language development - that is to say to the very first weeks of life - it becomes clear that this must be the case. The human infant is preeminently a maker of meanings, actively seeking to make sense of his experience in order to gain control over his environment. If we observe very young babies, we find that they have considerable skill in selectively responding to the regularities in their environment and in constructing and testing through action what can best be described as hypotheses about the organization of that environment. For example, Bower (1974) describes an experiment in which babies watch a steadily moving object and track it with their eyes. However, even as early as a few weeks, if the moving object is made to disappear behind a screen, babies are able to follow the projected trajectory and are already waiting when the object emerges from behind the screen. A considerable feat of learning and one that has involved no teaching at all. But this is only one of several experiments in recent years which demonstrate just how sophisticated young infants are in making sense of their environment.

However, the infant's environment is not purely physical. Even more important for his development is the fact that he is born into a social, actively responsive, environment. From the beginning, parents treat their babies as if they had intentions and as if their behaviour was intended to communicate those intentions. Research with mothers and infants as young as a few weeks shows that mothers interpret their infants' gestures and select and time their own contributions so that what is jointly achieved is something that looks like, and is experienced as, a conversation without words (Stern, 1977; Trevarthen, 1979). The explanation of this easy achievement of what has come to be called 'intersubjectivity' seems to be that, on the one hand, babies are born with a predisposition to produce rudimentary communicative behaviours and that, on the other, adults are equally predisposed to attend to such behaviours and to treat them as potentially communicative. The result is that, by having their behaviour selectively responded to in terms of the culturally significant intentions that are ascribed to them, infants do come to have and to express those intentions (Newson, 1978). That is to say that, by engaging in interactions with other people in which their gestures and, later, their vocalizations are treated as meaningful, they do indeed 'learn how to mean' (Halliday, 1975a). And, in the process, they construct a progressively more and more adult-like inner representation of the linguistic system that provides the culturally conventional means for their communication.

The evidence for children's continuing active construction of their linguistic resources has already been touched on. This is further confirmed, on the one hand, by the systematic nature of their successive transitional grammars which give rise to 'incorrect' utterances, which are never heard in the speech addressed to them (Bowerman, 1982) and, on the other, by the fact that in cultures in which adults adopt quite different practices with respect to the input that they provide, language learning proceeds in very similar ways (Heath, 1983; Schieffelin, 1979). Nevertheless, as already stated, this does not mean that the input is of no importance. To understand just how it contributes to language learning, however, we need to look more closely at the nature of conversation.

Although we often think of the goal of conversation as being that of exchanging thoughts and feelings, it is clear that, in any literal sense, such a goal is impossible. Thoughts and feelings are mental events that are unique to individuals; they arise in relation to the internal model of the world that the individual has constructed on the basis of his uniquely personal experience. Because no two people have had identical experience, their internal models are different and so their thoughts and feelings cannot achieve more than a partial correspondence. Furthermore, they cannot be transmitted direct from one mind to another but must be communicated by means of a linguistic code, in their command of which, again, individuals differ (Fillmore, 1979). As a result, rather than being a process of precise information transfer, conversation is, instead, a collaborative activity in which speaker and listener respectively provide and make use of linguistic and paralinguistic cues in context in order to enable the listener to construct an interpretation of what he judges to be the meaning intended by the speaker. As participants exchange roles in conversation, each speaker in turn builds on his version of the 'shared' meaning that has been established and, in the process, provides the previous speaker with feedback on how his own contribution was interpreted. The construction of conversational meaning is thus a cumulative and collaborative activity, which involves a considerable amount of implicit, and sometimes explicit, negotiation.

If this is true of conversation between mature adults who have similar biographies, how much more must there be collaboration when one of the participants is a young child, whose internal model of the world is both limited and idiosyncratic and whose linguistic resources are rudimentary in the extreme. Indeed, for the child to participate in conversation at all, the adult has to make considerable adjustments in order to compensate for the child's severe limitations. However, the adjustments called for are only extensions of those that are required for any successful and mutually satisfying conversation and, to varying degrees, all adults habitually modify their behaviour in these ways when interacting with young children.

These modifications include:

a. establishing and maintaining intersubjectivity of attention both by non-verbal behaviour such as posture and gesture and by exaggerated use of prosodic and paralinguistic features of speech (Bridges, 1979; Garnica, 1977).
b. treating the child's communicative behaviour as potentially meaningful and attempting to arrive at a plausible interpretation of it in context (Brown, 1973).
c. attempting to ensure, by means of checking expansions and reformulations (Brown, et al. 1969; Cross, 1977) that the meaning ascribed corresponds to the meaning intended by the child.
d. incorporating some aspect of the child's previous conversational contribution, either by extending the child's meaning or by inviting him to do so himself (Cross, 1977; Barnes et al., 1983).
e. selecting both the content and form of the response so that the adult utterance is appropriate to the child's current ability to comprehend (Cross, 1977, Snow, 1977; Furrow et al., 1979).

It is important to emphasize that parents vary in the extent to which they adopt these various conversational strategies. Whilst almost all adults in western cultures modify their non-verbal behaviour to secure their children's attention and adjust their speech in the direction of formal simplicity and semantic and contextual redundancy (the first and last of the strategies listed above), they show much greater variation in their ability or willingness to adopt the other strategies. Perhaps this is because (a) and (e) are necessary to ensure the child's take-up of adult attempts to achieve verbal control of his behaviour, whereas (b) - (d) only become important when the adult's aim is to engage collaboratively with the child in conversation.

Whether this is the conscious aim or not, the actual effect of the adoption of the full range of these strategies is to facilitate the child's participation in conversation and to provide him with clear evidence on which to base his construction of the language system and with feedback on the success with which he is able to use that system to communicate his own meaning intentions. At the same time, because such conversations are likely to be both relatively effectice in achieving the participants' aims and thus also mutually rewarding experiences, the child's motivation to engage in conversation is enhanced (Howe, 1981). As a result, he is likely to initiate conversation more often, with consequent advantages in terms of the increased opportunities to encounter evidence for his construction of the language system. Not surprisingly, it is children who experience a greater proportion of conversations of this kind who have been found to progress more rapidly in their language learning (Wells, in press a).

All these adult strategies are well illustrated in the following extract from a conversation between a 25 month-old boy and his mother. Mark was standing by the central heating radiator at the beginning of the extract and wanted to share his interest in his perception of heat with his mother.

Mark: 'ot Mummy(v)?	
	Mother: Hot? (checking) Yes that's the radiator
Mark: Been? Burn?	
	Mother: Burn? (checking)
Mark: Yes	
	Mother: Yes you know it'll burn don't you? (a few seconds later)
Mark: A man er dig down there	[looking out of the window]
	Mother: A man walked down there? (checking)
Mark: Yes	
	Mother: Oh yes
Mark: Oh yes	
	(a few seconds later)
Mark: A man's fire Mummy(v)	
	Mother: Mm?
Mark: A man's fire	
	Mother: Mummy's flower (checking)
Mark: No	
	Mother: What?
Mark: The man...fire (emphasizing each word separately)	
	Mother: Man's fire? (checking)

Mark: Yeh

Mother: Oh yes the bonfire

Mark: Bonfire

Mother: Mm

Mark: Bonfire
Bon-bonfire
Oh hot Mummy(v)

Bruner (1981) refers to such conversationally facilitating behaviour by adults as 'scaffolding', and he points out that, in the early stages, this requires the adult to assume a very much greater share of the responsibility for keeping the conversation going. But it is important to emphasize that, as in the very earliest pre-verbal interactions, this is most effectively achieved by letting the child initiate the conversational topic and making the adult contributions contingently responsive to those of the child. In fact, as Bruner notes, one of the characteristics of facilitating adults is that, once the child has gained some control of the linguistic resources, they relinquish the more dominant role and allow the child to be the conversational leader. In our longitudinal data, we found that children quite quickly reached the stage of initiating about two thirds of all conversations and that it was these child-initiated conversations that were the most likely to show real thematic development (Wells, in press a).

In sum, the role of the input is to provide evidence for the child's relatively autonomous 'reinvention' of the language of his community and to 'guide' that process by sustaining and extending the child's endeavours (Lock, 1980). The progressive approximation of the child's language system (his grammar) to that of the adult community occurs because he is predisposed to treat language as 'an internal problem space _per se_' (Deutsch, 1981) rather than because adults deliberately teach or systematically reinforce specific linguistic forms. In addition to providing evidence, however, the input can significantly facilitate the child's learning task if it is responsive to the child's output, that is to his speech and to his non-verbal indications of comprehension. Adults help children to learn if they treat their conversational contributions as worthy of careful attention and attempt to include them as equal partners in the collaborative construction of meaning.

LEARNING THROUGH LANGUAGE

So far we have been concerned only with language learning. But, at least for young children, learning language and learning through language are simultaneous and inter-related activities. With a few exceptions such as greetings, whenever people talk, they talk about some topic. In so doing, they collaborate in sustaining and mutually validating their cultural world-view and in calibrating their internal models of that world. For young children, therefore, conversation with other people, particularly with adults, provides evidence not only about the language system but also about the cultural view of the world that that language system encodes. Through conversation, as Halliday (1975b) puts it, children 'talk their way in', making an understanding of the adult world their own, as they construct their own mental models, adding to and modifying them in the light of their interpretations, in context, of the utterances that they hear.

Note the word 'interpretation', however. As already emphasized participants in conversation cannot communicate their thoughts directly to each other. The listener must construct a meaning intention to match what he believes the speaker intends and he does this by bringing the relevant aspects of his own mental model to bear on the cues provided by the speaker's utterance in context. Since the child's mental model is, by virtue of his limited experience, quite radically different in content and organization from that of the adult, so will be the interpretations that the child, as listener, is able to form of the adult's meaning intentions. An adult cannot, therefore, by telling, cause a child to come to understand the world as he understands it. He can only provide cues that he believes will enable the child to construct an interpretation that will allow him to achieve a more effective understanding of the matter in question. When the child replies, in his turn, the adult will then have an opportunity to form a more precise estimate of what the child's understanding is, and this, in turn, can form the basis for the adult's next contribution.

In general, though, parents are not concerned to instruct their children in any systematic way. With the exception of certain forms of acceptable social behaviour, they do not have specific matters that they want their children to learn and in no home that I know of is there anything approaching a curriculum. Learning is, therefore, for the most part incidental (Wood, 1983).

Yet children's learning is not entirely unsystematic. Whilst the particular items of information that they have at their disposal at any stage owe a great deal to the environment in which they are growing up and, in particular, to those aspects that are given salience in the conversations that they experience, the types of information that they are able to cope with and the ways in which they are able to organize that information are largely governed by the sequence of cognitive development, which is common to children whatever their cultural environment (Piaget, 1952, Cole and Scribner, 1974). Within these constraints, learning occurs most readily in relation to the purposeful activities in which the child engages and is built up around the topics that capture his interest.

Here again, though, it does not follow that, because adults can have only a limited influence on what the child learns and on how he learns it, their contributions are of little significance. Incipient interests can be encouraged and new ones suggested; those already apparent can also be sustained and extended. However, because the adult's contribution is made largely through the medium of conversation, facilitation of the child's learning will be best achieved by means of exactly the same strategies on the part of the adult as were found to facilitate the learning of language. That is to say, facilitation of learning through language equally requires collaboration and negotiation.

These qualities are clearly apparent in the following extract, in which, as his mother is trying to get him to change his shoes and socks, James's interest is captured by a bird in the garden.

	M: There - one slipper on
J: I can see a bird	
	M: A what love?
J: See a bird	
	M: Is there? (whispers) Outside?
J: Yes (whispers) See	[J points to bird]
	M: Is he eating anything? (whispers)
J: No (whispers)	
	M: Where? (whispers) Oh yes he's getting - Do you know what he's doing?
J: No (whispers)	
	M: He's going to the - the - paper sack to try and pick out some pieces - Oh he's got some food there And I expect he'll pick out some pieces of thread from the sack to go and make his nest... up ... underneath the roof James (whispers throughout)
	M: Wait a minute and I'll - OK wait a mo' wait a mo' James ...
J: That bird's gone (whispers)	
	M: Has it gone now?
J: Yes	
	M: Oh Take those long trousers off because they're... a bit muddy in there
J: Yes he's gone	

What is notable about this example is the way in which, quite intuitively, the mother responds to the child's interest with relevant information, not because she intends to teach him, but because she shares his interest. In the light of her own understanding of the bird's activities, she offers James the information that she thinks will be of most interest to him and most accessible to his understanding. As a result, James has the opportunity to enrich his initial understanding by working on the information contained in his mother's explanation. At the same time, it is worth noting how the quite different purposes of the two participants are negotiated. The mother drops her insistence on getting shoes and socks changed while she responds to the child's interest, but, once the bird has flown away, she returns to the purpose which is at the top of her own agenda.

Parents vary in their sensitivity to their children's interests, of course, and in the extent to which they are willing and able to sustain and extend them in a way that facilitates the development of their children's understanding. But, to varying degrees, the vast majority of children receive an input in the early years at home that not only provides evidence for the active construction of their mental models but does so in a way that is well adapted to their natural learning strategies. And this occurs, not because parents have clearly defined and sequentially structured teaching objectives, but because their conversational contributions are for the most part contingently responsive to their children's initiations.

How much children have learned by the time they enter school can be partially gleaned from overhearing their talk with their parents and with other familiar adults - as we did in the recordings that we made in their homes. These recordings contain many instances of quite sophisticated thinking and of equally sophisticated language. But this, it could be argued, is the result of a collaborative enterprise for which the adult shares the responsibility. What are they able to do on their own? To answer that question, we need to turn to talk in which only children are involved.

Peer-group talk is probably even more varied than talk with an adult. Children together joke, tease, quarrel and complain as well as ask and answer questions and generally use language for what Halliday (1975a) calls 'heuristic' and 'informative' purposes. But probably the fullest use they make of their linguistic resources is in cooperative dramatic play. Here, instead of utterances being selected to fit the real world in order to request, or inform in relation to real events, language is used to create an alternative possible world - an imaginary world of fantasy, in which they take on and abandon roles as the requirements of the developing story demand. Eavesdropping on this sort of talk confirms the view that, by the age of five, children have become quite skilful at dealing with relationships of motive, cause and consequence in their thinking and in their expression of them in talk. What is perhaps more surprising however, is to discover how skilful they also are at managing their interpersonal relations as they interpret each others' intentions and negotiate a jointly constructed story world.

In the following extract, Sam, John and David are playing with a varied set of playpeople and animals. David has a cardboard box; this is his 'base', in which he is arranging some of the playpeople and their possessions. Sam also has his own territory, a wooden boat, on which he has a family of lions. All around is the sea - the playroom carpet. At this point in the play, John, who also has a boat and an assortment of playpeople, is torn between joining David on his base or Sam on his boat. The problem is that neither base nor boat has room for all John's people. To a large extent, these preoccupations with territory and space provide the underlying motivation for the way in which the drama develops. However, it is the imaginary characteristics and needs of the playpeople and animals of the story world which give the drama its surface texture. (Utterances in italics are spoken in 'play' voices appropriate to the characters concerned).

Sam	John	David
1 Now you have to live on your boat (to J, to take his play people somewhere else)		(to self) 'Tend it was put down like that (arranging his base)
	2 Why? Why do we?	
3 Cos there was no room for you (i.e. on the lions' boat)		(D puts people, furniture etc. into his base)
	4 Pretend we was sending boats back (moves S's boat with the lions on)	
5. *No that's our boat that's our boat* (lions speaking)		
	6 No but pretend we was sav- saving them back so people could get um-	
	7 That was your fault (to D who has got in his way)	
8 OK we're living on here (i.e. on base) Oh we'll die		
	(begins to put his people on D's base)	
		9 We- we've got all the luggage
		10 *I'm going to sleep (pretends to cry)*

11 All our luggage is- is-

12 One of er- one of our boy-friends is crying in a corner (pretends to cry)

13 Pretend one of the- the - their children was crying in a corner (pretends to cry)

14 Why was that?

15 It was because they didn't like being on the - (pretends to cry)

16 They didn't like being on land

17 -All squashed up did they? (i.e. on the base)

18 No they didn't like being-

19 They went outside didn't they?

20 Yeh and they had to go out

21 And it was poison on the sea and they had to die didn't they?

22 No they didn't They got on this boat (i.e. the lions' boat)

23 They jumped on to there They was good jumpers

This sort of talk is very different from that in the preceeding examples. Its coherence is much less apparent and indeed the thread is quite difficult for an outsider to follow. Not for the children, however. They are obviously adept at simultaneously negotiating the inter-personal relationships among themselves and developing the story line. Sam defends his territory against John (3) with a justification in terms of the lions' needs and, when John makes a counter-suggestion, he has the lions speak in their own defence. However, at this point, David introduces a new theme, which breaks the impasse. Alternately constructing the story and acting the part of one of the characters, he develops the theme of overcrowding (10-17), prompted by Sam's question (14). Sam, in turn, extends this with a further theme - death by poison (21), which John rejects in favour of a less drastic result (22).

The point of this example, then, is to open another window on the world of childhood and on children's talking and thinking. As

Phillips (in press) points out, when an adult is present, the talk tends towards an adult view of the world. Indeed, as already argued, it is through such talk that children engage in the progressive construction of that adult reality. However, they also inhabit another reality, which is closer to myth, fairy tale and magic. Most of us as adults, have left this world behind and can probably no longer remember it - at least not from the inside. Because it is no longer a part of our thinking, though, does not mean that it is not important for children. However we choose to respond to it, it is part of what they bring to the learning-teaching encounter.

LANGUAGE, LEARNING AND TEACHING IN SCHOOL

From observations of children at home in the pre-school years, we gain a powerful impression of them as active makers of meaning, constructing their linguistic resources and, simultaneously, as they use those resources to converse with those around them, constructing an internal model of the world in which they live. Adults have an essential part to play in providing the evidence that children need to engage in these processes, but they do not control the sequence in which learning takes place nor what the outcome of that learning will be. These rest with the child and result from an interaction between his innately given sense-making procedures and the evidence to which he is exposed. The adult's contributions are most helpful to the learner, therefore, if they are offered in response to his initiations rather than imposed in a didactic fashion. It is precisely this sort of facilitation that Vygotsky (1962) seems to have had in mind when he urged the importance of the adult working with the child in the child's 'zone of proximal development'. Lock (1980) describes this collaboration between child and adult in the language-learning process as 'the guided reinvention of language'. By extension, it seems equally appropriate to describe the learning about the world that takes place through adult-child conversation as 'the guiged reinvention of knowledge'.

If this is indeed the most appropriate way to conceptualize the relationship between child and adult - between learner and teacher - there are quite radical implications for the way we think about schools and classrooms. Perhaps the most important of these is that it gives a new significance to the well-known precept to teachers: <u>start where the child is</u>. All too often this is interpreted in practice to mean: administer a test or some other form of assessment in order to decide which ability group to place the child in or which reading primer or work-sheet to give him. But this is not discovering where the child is - what his concerns, needs and interests are. Instead, it is discovering into which of the places that we have prepared in advance he can most easily be slotted. Really to discover where a child is and, hence, how we can most helpfully contribute to his further learning, it is necessary to listen to what he or she has to say - to try to understand the world as he or she sees it. Only then can the teacher's contribution have that quality of contingent responsiveness that we have seen to be essential in helping the child to extend and develop his or her understanding.

A further implication of recognizing children's active involvement in their learning is the importance of encouraging them to share the responsibility for deciding what tasks to undertake and how to set about them. Effective learning is only marginally concerned with coming to be able to demonstrate skills and to recall facts on the demand of others. Much more important is being able to draw upon skills and to select and use information to solve problems that have meaning for the learner because he has taken them on and made them his own. Seven-year-old Colin, in the following extract, has a real problem to solve as he tries to make a tripod to support his model camera. In this situation, his teacher's suggestion that he consults reference books and tries drawing a scale plan is received as helpful advice and these activities have a genuine purpose for him rather than, as so often, being undertaken as arbitrary teacher-imposed tasks.

T: Coling are you having a
problem?

C: Just trying to ...
think out... something
Just trying to think out
how high I want the pole

T: Could you work there a while
(to another child)
I'll just help Colin [T joins C, who is using a metre rule, extending it with a small ruler. He is reading off the height he needs.]

C: One metre and -

T: Can you imagine for a minute
that you 're taking a photograph
How high would be comfortable?

C: Er- this is what I done -
trying to find that out
I put this like that and
held it and just pretend
that I was looking through
and I thought I'd have it
about that high cos that
could include the camera
on top
And that's how far I want
it - one metre and thirteen
centimetres [C counts on small ruler]

T: Is that going to be the height of your tripod

C: Yes - of the pole

T: Is each pole going to be that height?

C: I'm only going to have three um - yes
The other two are going to be a bit longer

T: Can you show me how you're going to do your plan?

C: I've got a lump of wood

T: Pardon?

C: I've got some wood and that's what it's going to look like [C points to his plan]
It's going to have those bits so I can put something around it to hold the camera on and...
I'm going to try and get something that can - a round hole - but that could hold on ... on legs which is going to be rather hard

T: Have you looked in the camera book to see if it shows a diagram that would help you?

C: Er - I have looked in one

T: Did you notice that there was another one there today?

C: No - yes there is

T: Perhaps in a moment you'd like to look at that
That might be helpful

C: Yes

T: What else will you need?

C: Um yes - a sharp tool that I can make the ends of them... rather sharp so they can dig in the ground
Or I could have blunt ones that just stand out to keep it steady

T: And how will you set your tripod up?

C: It's going to be always set up (laughs)
Just all you have to do is
Just take it outside to something ... like that

T: How do you think that's going to improve your photography?

C: It's going to keep it much stiller and the pictures will be much better ... cos they won't go blurry through movement

T: Will you have a look at that book with me?

C: Yes [C settles down to look at reference book]

Our observations, unfortunately, show that this sort of collaborative approach to learning in school is rare. Most of the tasks that children engage in are unilaterally imposed by the teacher and are selected in terms of a curriculum that has been worked out in advance from the adult perspective of the goals to be reached and the route to be taken to reach them. And because young children are anscious to please their teachers, they fairly readily engage in these tasks, although they may have little intrinsic significance for them. But what do they learn in the process? They may, for example, acquire the 'component skills' of reading, but will they go on to read books outside the classroom for pleasure or to obtain information? They may acquire an assortment of items of information, but will these be incorporated into their personal mental models of the world so that they become knowledge that can be drawn upon to guide their own undertakings? Donaldson (1978) argues that activities which are undertaken in order to achieve external reward are less likely to be undertaken later when the reward is absent; they are also less likely to be enjoyed.

Just as children typically have little active involvement in deciding on the learning tasks in which they engage, so they also seem to have a rather passive role in the teacher-pupil talk in which these tasks are embedded. As part of our longitudinal study, we made a direct comparison of the language experienced by children in the two settings of home and school. Seven 5-minute samples, taken at 20-minute intervals between 9 a.m. and 12 noon, were recorded in both settings and all adult-child interaction was submitted to a comprehensive analysis (Wells, in press b). Table 1 summarises the result of this analysis.

As can be seen, compared with the situation at home, these children played a much less active part in conversation with an adult at school. They initiated significantly fewer interactions, were allowed fewer turns per interaction and, proportionally, asked fewer questions and made fewer requests. Furthermore, their individual utterances were on average syntactically less complex and contained a narrower range of meanings. Indeed, a

Table 1. Comparison of Adult-Child Conversation at Home and School (n=32)

	Home	School	Sig. level of difference
Absolute values			
Mean no. of child utterances to adults	122.0	45.0	p< .001
Mean no. of adult utterances to child	152.7	128.7	n.s.
Mean no. of child turns per interaction	4.1	2.5	p< .001
Mean child syntactic complexity	3.1	2.4	p< .001
Mean adult syntactic complexity	3.5	4.3	p< .001
Mean no. of categories of semantic content in child speech+	15.5	7.9	p< .001
Proportional values (child)			
Initiation of interaction	63.6%	20.3%	p< .001
Exchange-initiation utterances	70.2%	43.8%	p< .001
Complete statements	31.2%	28.0%	n.s.
Questions	12.7%	4.0%	p< .001
Requests	14.3%	10.4%	p< .05
Elliptical of moodless utterances	29.4%	49.4%	p< .001
Utterances in text-contingent exchanges	9.4%	6.3%	p< .10
References to non-present time	9.1%	6.4%	p< .05
Proportional values (adult)			
Exchange initiation utterances	59.9%	78.7%	p< .001
Complete statements	26.2%	24.5%	n.s.
Questions	14.3%	20.2%	p< .01
Requests	22.5%	34.1%	p< .001
Elliptical utterances	5.7%	5.8%	n.s.
Requests for display	2.1%	14.2%	p< .001
Extending child's meaning	33.5%	17.1%	p< .001
Developing adult's meaning	19.3%	38.6%	p< .001

+ For this comparison only n = 16

much higher proportion consisted only of sentence fragments, occuring frequently as minimal responses to requests for display.

These differences in the children's language use were the fairly direct result of differences in the adults' behaviour in the two settings: for, compared with the parents, the teachers tended to dominate the conversations, initiating the majority of interactions and maintaining their controlling role through questions, requests and requests for display. However, the most significant difference between the two settings was to be seen in the differing proportions of adult utterances that extended the children's meanings. Compared with the parents, the teachers were far more concerned to develop their own topics than to accept and develop the topics offered by the children. The result was that, at school, the children actually had less opportunity to learn through talk with an informed adult than they did at home. Nor war this the case only for the most able children. Even the linguistically least advanced were encouraged to make greater use of their linguistic resources at home than they were at school. In fact, it was precisely the least able children who suffered most. For, in teacher-dominated question and answer sequences, these children were frequently reduced to apparent linguistic incompetence, thereby confirming their teachers' low expectations of them (Wells, 1982). And, because such children are frequently assumed to be able to cope best with a highly-structured skill-based type of instruction, they were likely to have even less opportunity than their peers to put their already proved active learning strategies to effective use in school.

Since these results were almost certainly not what the teachers intended, it is worth asking what it is that leads teachers to behave in ways that, all too often, have as a consequence an actual reduction in children's opportunities for effective learning as they make the transition from home to school.

To some extent the reasons have already been suggested: the difficulty of having charge of such large numbers of children and the increasing pressures from outside the classroom to show that all the children for whom they are responsible are systematically covering the curriculum. Under such conditions, it is not surprising that many teachers find themselves reverting to a 'transmission' model of teaching, particularly as their own experience as pupils and students was very probably of this kind.

Nevertheless, as I have tried to show, the transmission model can never be really effective, because it ignores the contribution of the learner and, indeed, the very nature of the learning process, as this takes place through interaction with another person. The teacher's knowledge or understanding can never be transmitted to a pupil on the teacher's terms, for knowledge and understanding can only be constructed by each individual for him- or herself, starting from where he or she is. Learning is an active process in which the learner modifies his internal model of the world in response to the interpretation he is able to put upon the new information that is presented to him. But that interpretation, in turn, depends not only on the clarity with which it is presented, but also on the current state of the learner's internal model, as a result of previous experience, and on the extent to which he is motivated by having a purpose for learning which is important to him. Learning is also an interactive process, both in the Piagetian sense just described of accommodation and assimilation between internal model and new information, and also in the requirement for collaboration by teacher and learner together in the joint construction of meaning.

Some pupils are certainly able to learn with some success from a style of teaching characterized by prescribed tasks, instruction, questioning and evaluation, delivered from the teacher's perspective. But others, equally certainly, are much less successful. It is my conviction, however, that <u>all</u> pupils learn more successfully if teaching is seen not as the transmission of predetermined knowledge possessed by the teacher, but as the facilitation of individual pupils' learning. This requires precisely the same sort of collaborative meaning-making that was so successful in the preschool years in enabling children simultaneously to learn to talk and to learn through talk.

Nevertheless, as has been repeatedly stressed throughout this paper, such a conception of the relationship between learning and teaching does not in any way diminish the importance of the teacher's contribution. Pupils need guidance if they are to learn effectively: they are helped by suggestions as to fruitful lines to follow and by discussions that enable them to evaluate their success in achieving their goals. However, this guidance will be all the more effective if it is contingently responsive to pupils' own efforts to attain goals that they have taken on and made their own.

The most successful teachers have always known this and have endeavoured, themselves, to be models to their pupils of effective learners. In various ways they have provided opportunities in their classrooms for their pupils to learn alongside them in a master-apprentice relationship. Many of us will have been fortunate enough to have met teachers of this kind in our careers as learners and almost all of us will at some time have experienced the excitement of learning from a project undertaken from our own choice and the satisfaction of meeting the challenge to communicate our resulting expertise to others.

As teachers, we need to remember these successful learning experiences and to find ways of organizing our own interactions with pupils so that they, too, have similar opportunities. However, there is no simple formula for achieving this sort of collaboration in learning. Different topics call for different approaches, as do the different stages of development, backgrounds and personalities of pupils and teachers (Wells, 1978). Finding satisfactory patterns of classroom organization to make possible these different approaches calls for a willingness to experiment and to involve pupils in sharing the responsibility for making them work. The support of colleagues is also a great help. Where several teachers in a school, or still better the whole staff, regularly meet together to share their experiences of approaches that have worked and to discuss alternative possibilities, a momentum will be built up that can carry an individual teacher through an occasion when things go wrong.

The important step is the first one. For once teachers begin to engage in collaborative meaning-making with their pupils, the experience will almost certainly be such as to make them wish to continue. To any teacher who needs to be convinced that such a step is worth taking, I would simply suggest that they taperecord and transcribe an hour or two of the talk that occurs in their own classroom. This will almost certainly convince them that they can teach more effectively when they think less about their prepared input and concentrate more on helping pupils to make the knowledge their own.

REFERENCES

Barnes, S.B., Gutfreund, M., Satterly, D.J, and Wells, C.G., 1983, Characteristics of adult speech which predict children's language development. Journal of Child Language, 10: 65-84.

Bower, T.G.R., 1974, Development in Infancy. San Francisco: Freeman.

Bowerman, M., 1982, Reorganizational processes in language development. In Wanner, E. and Gleitman, L.R., eds., Language Acquisition: the State of the Art. Cambridge: Cambridge University Press.

Bridges, A., 1979, Directing two-year-old's attention some clues to understanding. Journal of Child Language, 6: 211-226. Cambridge University Press.

Brown, R., 1973, A First Language: the Early Stages, London, G. Allen and Unwin.

Brown, R., Cazden, C. and Bellugi, U., 1969, The child's grammar from I to III. In Hill, J.P., ed., The 1967 Minnesota Symposium on Child Psychology, Vol. 2. Minneapolis: University Minn. Press.

Bruner, J.S., 1981, The pragmatics of acquisition. In Deutsch,W., ed., The Child's Construction of Language. London: Academic Press Cambridge, Mass.: M.I.T. Press.

Chomsky, N.A., 1965, Aspects of the Theory of Syntax, Cambridge, Mass: M.I.T. Press.

Chomsky, N.A., 1976, Reflections on Language London: Temple Smith

Cole, M. and Scribner, S., 1974, Culture and Thought: a psychological introduction. New York: Wiley.

Cross, T.G., 1977, Mother's speech adjustments; the contribution of selected child listener variables. In Snow, V. and Ferguson, C., eds., Talking to Children, Cambridge: Cambridge University Press.

Deutsch, W., 1981, Introduction. In Deutsch, W., ed., The Child's Construction of Language London: Academic Press.

Donaldson, M., 1978, Children's Minds, London Fontana.

Fillmore, C.J., 1979, On Fluency. In Fillmore, C.J., Kempler, D. and Wang, W. S-Y, eds., Individual Differences in Language Ability and Language Behavior New York: Academic Press.
Furrow, D., Nelson, K., and Benedict, H., 1979, Mother's speech to children and syntactic development: some simple relationships. Journal of Child Language, 6: 423-442.
Garnica, O.K., 1977, Some prosodic and paralinguistic features of speech in young children. In Snow, C.E. & Ferguson, C., eds., Talking to Children: Language Input and Acquisition, Cambridge: Cambridge University Press.
Halliday, M.A.K., 1975a, Learning How to Mean London: Arnold.
Halliday, M.A.K., 1975b, Talking one's way in: a sociolinguistic perspective on language and learning. In Davies, A., ed., Problems of Language and Learning. London: Heinemann/S.S.R.C.
Heath, S.B., 1983, Ways with Words, Cambridge: Cambridge University Press.
Howe, C., 1981, Acquiring Language in a Conversational Context, London: Academic Press.
Lock, A., 1980, The Guided Reinvention of Language, London: Academic Press.
Moerk, E.L., 1983, The Mother of Eve - as a First Language Teacher. Norwood, N.J., Ablex.
Newson, J., 1978, Dialogue and development. In Lock, A., ed., Action, Gesture and Symbol London. Academic Press.
Phillips, T., in press, 'Talk among yourselves: it's not my style'. In Wells, C.G. and Nicholls, J.C., eds., Language Development and Education. Brighton: Falmer Press.
Piaget, J., 1952, The Origins of Intelligence in Children. New York: International Universities Press.
Schieffelin, B.B., 1979, How Kaluli Children Learn what to say, what to do and how to feel; an ethnographic study of the development of communicative competence. Unpublished doctoral dissertation, Columbia University.
Skinner, B.F., 1957, Verbal Behaviour, New York: Appleton.
Snow, C., 1977, Mother's speech research; from input to acquisition. In Snow, C. & Ferguson, C., eds., Talking to Children. Cambridge: Cambridge University Press.
Stern, D., 1977, The First Relationship: Infant and Mother. London: Open Books.
Trevarthen, C., 1979, Communication and Cooperation in Early Infancy A Description of Primary Intersubjectivity. In Margaret Bullowa, ed., Before Speech: The Beginnings of Human Communication. Cambridge: Cambridge University Press.
Vygotsky, L.S., 1962, Thought and Language Cambridge, Mass. M.I.T. Press .
Wanner, E. & Gleitman, L.R., eds., 1982, Language Acquisition: the State of the Art. Cambridge: Cambridge University Press.
Wells, C.G., 1978, Talking with children: the complementary roles of parents and teachers. English in Education, 12, 2:15-38. Reprinted in Language, Learning and Education. Centre for

the study of Language and Communication, University of Bristol, 1982.

Wells, C.G., 1981a, Learning through Interaction: the study of Language Development, Cambridge: Cambridge University Press.

Wells, C.G., 1981b, Some antecedents of early educational attainment, British Journal of Sociology of Education, 2: 181-200.

Wells, C.G., 1982, Language and learning. In Language, Learning and Education. Centre for the study of Language and Communication, University of Bristol, 1982.

Wells, C.G., in press a, Language Development in the Pre-school Years. Cambridge: Cambridge University Press.

Wells, C.G., in press b, The language experience of five-year-old children at home and at school. In Cook-Gumperz, J., ed., Literacy, Language and Schooling. Exeter, New Hampshire: Heinemann Educational.

Wood, D., 1983, Teaching: natural and contrived. Child Development Society Newsletter, 31: 2-7.

NATIVE AND SECOND LANGUAGE ACQUISITION AND LEARNING

CAPACITY-SHARING INTERDEPENDENCE IN READING PROCESSES

M. Boekaerts

Dept. of Psychology and Educational Research

University of Nijmegen

1.0 INTRODUCTION

Psychologists have studied the reading process for many years. The main reason for their persistent interest in the reading act undoubtedly lies in the fact that reading is a complex skill which involves many interacting subprocesses. Two main approaches can be distinguished : (1) a pratical approach and (2) an explorative approach. The rationale behind the former approach is to identify ways to improve instruction. On the basis of cognitive principles experimental instructional programmes are designed and they are tested in a natural or quasi-natural setting. If they prove to be more successful than the traditional reading programmes they are implemented and henceforth replace the existing programmes. By contrast, researchers who foster the latter approach have as their main objective the analysis and description of specific aspects of the reading act. They adhere to a specific reading model, featuring one or more specific subskills, and try and expand our knowledge of those subskills by way of controlled experiments and hypothesis testing.

The decision to focus on specific subskills is based on the practical principle that a 'manageable' subset of the processes involved in the reading act should be studied. The advantage of such careful and systematic analysis is that detailed descriptions can be given of the rules governing specific procedures and strategies (e.g. Perfetti's model describing decoding strategies). An important draw-back of dissecting the reading process into component skills is that the subprocesses are studied in isolation and that the interaction between the subprocesses is overlooked. This may lead to the description of 'artificially, generated or distorted subprocesses. Indeed, if we want to describe the actual reading process we should not forget that it is a multi-level processing task made up of a number of simultaneous and interacting subprocesses (Clark & Clark, 1977; Marslen-Wilson, 1975); Levelt, 1978; Rumelhart, 1977; Tanenhaus, 1978).

An alternative to the 'small-spectrum models' is the description of the complete reading act of good and poor readers (expert-novice comparisons). Several attempts have been made to build such comprehensive models of reading comprehension (Gough, 1972; Goodman, 1970; Rumelhart, 1977). Before discussing the advantages and shortcomings of these models, I would like to refer to Gibson's and Levin's (1975) warning that a single model of reading is counterproductive because it can be nothing more than a gross oversimplification of the real act. They point out that there are marked individual differences between the use of procedures and strategies and that these differences may be subject-matter-and goal-specific.

How seriously should we take this warning? Should we refrain from model-building altogether? Coltheart (1977) and many others argued that the advantage of constructing reading models is to be found in the fact that an author is forced to specify the sequence of operations that a reader must perform in order to read. As such, he must make explicit how the reader performs the various operations (i.e. the arrows in a flow-chart) and how he can consult the various sources of information (i.e. the boxes).

When a model is viewed as a detailed description of a series of systematic and finite steps that lead to a skilful reading performance, it is analogous to a computer programme. If it works, considerable progress has been made; if it doesn't work it shows up the gaps in knowledge and can direct future research. These gaps must be bridged by speculation or by assumptions based on incomplete evidence. As Gough (1972) pointed out information about the reading process can be accrued by describing the various subprocesses in great detail so that they can be proved either or incorrect.

To conclude, there have been many attempts to describe the subskills of the reading act. The results obtained on the basis of small-spectrum reading models are hard to compare in many respects and it is far from clear what their joint efficacy is at present. Severe criticism has also been levelled at most of the wide-spectrum reading models. It would take me too long to present and discuss these arguments here. I shall also refrain from presenting yet another conceptual framework depicting the reading process. (The interested reader is referred to excellent surveys such as Gibson & Levin, 1975; Guthrie, 1976 ; Mitchell, 1982; Singer & Ruddell, 1976). Instead, I shall list a number of generally accepted principles of reading that apply in a variety of reading situations. Subsequently, I shall focus on one important aspect of the reading process; namely the limited capacity of the central processor.

Twelve important principles concerning the reading act:

1. The reading act should not be seen as the execution of a complex algorithm. It is a highly complex, but flexible heuristic procedure.

2. In order to get at the meaning of a printed message, the reader can make use of three main types of analyses: (1) linguistic analyses at different levels of complexity (the letter, word, phrase, sentence, text level), (2) conceptual analyses or schema-driven analyses and (3) contextual analyses or making use of cues that are present in (put into) the reading material or the reading environment (e.g. pictures, titles).

3. A number of functionally different reading activities can be distinguished, e.g. reading aloud, silent reading, reading familiar or unfamiliar material, reading printed material, reading handwriting or degraded material, reading within a time limit, reading in the presence or absence of others, etc. The assumption is that the reading act is essentially the same in these various activities but that differences in the interaction between the subprocesses must be postulated.

4. Different readers may want to reach fundamentally different goals with the same reading activity. For example, they may want to prove that they can read technically or fluently; they may want to gain or retain information; they may want to skim through the material; they may want to read for pleasure. The reading goal may however alter the nature of the reading act.

5. The efficient steering of eye-movements across the line and over the page (forward movements and back-ups) is an essential subprocess of effective reading.

6. The total level of resources available for conscious information processing is limited.

7. The beginner's reading process is characterized by an overaccentuation of the decoding process (decoding individual letters and words). The high-capacity consuming, serial mode of information processing is constantly in use.

8. Poor readers compensate for less adequate basic processing skills (linguistic analyses at different levels) by making explicit use of the semantic context and information in long-term memory).

9. As the reader gains in decoding skill, basic processing -especially at the word level- requires less active processing capacity, leaving spare capacity for higher order processing.

10. In skilful readers the decoding process is a fully automatic subprocess of the total reading act. It is neither necessary to vocalize individual words nor to make reference to the semantic context.

11. When word recognition fails (due to unfamiliar, ambiguous or degraded words and phrases) skilful readers make use of various sources of information (language-specific rules such as orthographic rules, pronunciation rules, morphological rules, syntactic rules) to disambiguate the message.

12. Skilful readers use an efficient time-sharing principle. They can skip redundant information, execute subprocesses on 'automatic pilot' and leave enough processing capacity for higher order processing (interpretation, retention, etc.). When necessary they can focus on, or regress to functionally relevant ambiguous or emotionally valued sections of the text without disrupting the total reading act.

For the sake of convenience I have summarized present conceptualizations about the reading process into twelve summary statements. I have thereby emphasized two important facts: (1) that all subprocesses of the reading act interact and (2) that all subprocesses draw upon the same limited resources. The former of these facts refers to interactive theories of reading comprehension as proposed by e.g. Norman & Bobrow, 1975; Clark & Clark, 1977; Rumelhart, 1977; Stanovich, 1980. These authors argue that processing efficiency as well as processing rate are influenced by the nature of the interaction between the various subprocesses. For example, if the subprocesses at a particular level of linguistic analysis are hampered due to visual degrading, data-processing at other levels of analysis will be influenced. This type of data-sharing interdependence produces facilitating or inhibitory effects. These effects have been well documented (e.g. Samuels, 1970; Guthrie, 1973; Perfetti et al, 1979) and they shall not be treated in this article. Rather, I shall concentrate on the other form of interdependence, capacity-sharing interdependence.

LaBerge & Samuels (1974) put forward a theory of reading in which the concept of limited processing capacity holds the central position. They state that if two or more subprocesses make heavy demands on the joint capacity of the central processor, other subprocesses will be impaired or slowed down. In other words, they postulated capacity-sharing interdependence between the various subprocesses of the reading act. In this article I will further explore the concept of capacity-sharing interdependence. My main concern is to devise an experimental technique that is sensitive enough to determine whether the various subprocesses of the reading act have reached a satisfactory level of automaticity.

2.0 AUTOMATIC VS. CONTROLLED MODES OF INFORMATION PROCESSING

It is common-sense experience that when reading a not-too-difficult text one is able to execute a number of additional subprocesses such as for example 'thinking of going home early that day'; 'thinking of an example of the points mentioned in the text', etc. In other words, when processing demands are low, reading-specific subprocesses proceed on automatic pilot, leaving enough spare processing capacity for retention purposes and for processing task-relevant or task-irrelevant information. However, when task demands increase due to either an increase in difficulty level (new, unexpected or ambiguous words or phrases) or deteriorated reading conditions, the reader is forced to allot more processing capacity to the reading task per sé. Shiffrin & Shneider (1977) distinguished two modes of information processing in this respect: (1) the automatic mode and (2) the controlled mode. Reversals, or switches from one mode to the other occur spontaneously when task demands increase or decrease; for, the division of attention among various subskills is normal when performing complex tasks. However, when the reader is continuously forced to switch to the high-capacity consuming, serial mode of information processing in order to assist the decoding or comprehension process, the delicate balance between the various subprocesses is distorted. In such extreme cases an instant drop in reading performance will be noticed. When reversals are more subtle, a smooth reading performance may be interrupted temporarily and may hardly be noticed. These reversals may nonetheless be indicative of a less than adequate reading performance. In an educational setting this information may be very helpful in selecting adequate reading material. Indeed, it is not so difficult to identify either an excellent or a poor reading performance. But, it is hard to differentiate between the reading performance of two pupils, or the performance of a pupil on two different reading trials of we have neither a measure of the joint processing demands of the reading task nor of the spare capacity that is still available while performing it.

In order to determine how 'good' a 'good' performance really is, we need information about when and why reversals occur. Is it because of inefficient time-sharing among the various subprocesses (i.e. the level of automaticity of one or more subprocesses has not been reached yet) and/or is it due to input overload (difficulty level of the reading task/task conditions)? I believe that the best way to gain this information is to ask subjects to perform a secondary task simultaneously with the reading task. The dual task technique can provide an estimation of the processing demands of the combined subprocesses of the reading act. By systematically manipulating the complexity of the reading task the effects of the respective primary tasks on the subject's performance on the secondary task can be studied. The experiment reported below investigates whether an on-line, open-loop psychomotor task is a suitable secondary task for detecting switches from the automatic to the controlled mode. The dependent measures in this study are (1) the number of errors, (2) the number of disruptions in the secondary task and (3) the comprehension score. The major comparisons of interest with respect to these performance measures concern (a) the task-specific effect (reading task vs. reciting task) and (b) the linguistic complexity effect. We predict that there will be no task-specific effect but that increases in the linguistic complexity of the primary task will result in an increase in the number of disruptions in the secondary task.

3.0 METHOD

Subjects

Fourteen primary school pupils of the Sint-Stevenschool in Brussels (6th form 11-12 year-olds) volunteered to take part in the experiment. All the subjects were right-handed, had normal or corrected-to-normal vision and normal hearing. Each subject was taken individually by his/her own teacher in a total of 5 experimental sessions lasting about ten minutes each. The experimental sessions were incorporated -as much as possible- into normal classroom activities.

Apparatus and stimuli

The apparatus consisted of an adapted two-track tape recorder, a stop watch, a secondary task and 2 types of primary tasks. A pilot study was run to select the secondary task in connection with the demands of the primary tasks. We finally selected an open-loop psychomotor task (Keele, 1968; Boekaerts, 1984) that can easily be performed concurrently with the primary task (even by six-year-olds). This task consisted of holding up the left arm and rhytmically opening and closing the hand. With the aid of a small electronic device strapped to the left hand the repetitive hand movements were converted into an acoustic signal. These pip-tone pulses were recorded on the second track of the tape recorder so that they could later be superimposed on the singing, reciting or reading-aloud performance (first track).

As primary tasks we selected (1) highly overpractised reciting tasks (well-known rhymes and songs) and (2) familiar and unfamiliar reading texts. The familiar texts consisted of already practised reading material. As unfamiliar texts we selected the AVI-reading charts. These charts consist of 2 sets of 9 reading texts increasing in difficulty level (A and B-version). The difficulty level of a text is determined on the basis of its linguistic properties (word complexity and wordlength; sentence complexity and length of sentences; lay-out and graphemic presentation). For each text reading norms were available (reading time and error rate). We used the A-version to determine each pupil's reading level. The matched B-version was used for the experimental conditions.

Design and procedure

When subjects were able to perform the secondary task as a single task (usually after one trial) they received further practice in performing this psychomotor task simultaneously with several reciting tasks. When the secondary task could be performed without disruptions in either the reciting task or in the hand movements we proceeded with familiar reading texts. When we felt confident that all subjects could perform the two tasks simultaneously we collected base-line data (average number of pulses per one minute period) and started the experiment. It appeared that all 14 pupils were situated between difficulty index 7 and 9. Hence, we started at level 7 with the B-version. In all reading conditions subjects were instructed to read the text aloud (one-page put in front of them) for a general understanding and not to worry about the detailed information. After each reading trial they were tested for reading comprehension by a set of 3 multiple choice questions. They were not allowed to use the text but were given feedback on their answers. For all reading texts, reading time (RT) and errors (ER) were also measured. Errors consisted of (1) skipping one or more words, (2) reading words that were not present in the text and (3) spending more than 5 seconds to decode a word; in that case the teacher assisted the decoding process. Words or phrases that a pupil corrected himself, halts and pauses were not considered as errors.

As a measure for secondary task performance the number of scored irregularities (disruptions in timing) in the acoustic signal was used were counted as disruptions all shifts from a continuous rhythmical pulse to either a non-stop signal (i.e. the subject forgot to open his hand thus pressing the electronic device firmly) or silence (i.e. the subject forgot to close his hand).

4.0 RESULTS

Table 1 shows the mean number of errors, the mean number of disruptions in the secondary task and the mean number of multiple-choice items answered correctly. To discover whether the noted differences in these patterns of performance vary according to the type of dual-task, three separate ANOVA's (treatment by subject design) were conducted. Significant differences between conditions were found: (1) error $F= 13.34\ (4,65)$, $p < .0001$, (2) disruptions $F= 33.79\ (4,65)$, $p < .0001$, and (3) comprehension $F= 31.54\ (4,65)$, $p < .0001$.

In order to interpret these significant effects, three Newman-Keuls multiple comparisons (single factor with repeated measures) were run. As indicated in table 2 no significant effect was found between condition 1 (reciting while clapping) and condition 2 (reading a familiar text). There is however a significant-effect between condition 3, 4 and 5 (reading an unfamiliar text) and conditions 1 and 2. This difference shows up in the disruption and comprehension data but not always in the error data. The predicted linguistic complexity effect did not lead to a significant difference between conditions 3 and 4 and between conditions 4 and 5 but the difference was significant between conditions 3 and 5.

5.0 DISCUSSION

We predicted and found that there was a non-significant effect between a reciting task and a reading task per sé. We also predicted that the pupil's performance on both the primary task and the secondary task would be related to the linguistic complexity level of the primary task. Although the results of the ANOVAs indicate a linguistic complexity effect, the results of the multiple comparison tests do not warrant the conclusion that the conditions are mutually exclusive. For example, the difference between condition 3 (unfamiliar reading text at the level of the reader) and conditions 1 and 2 is not picked up in the error data although it is reflected in both the disruption and the comprehension scores. This result may lead to the conclusion that the pupils paid a lot of attention to the reading aloud performance (thus avoiding errors) but that the capacity needed to reach that performance

Table 1: Mean number of errors, mean number of disruptions and mean number of comprehension questions answered incorrectly.

	condition 1	condition 2	condition 3	condition 4	condition 5
mean number errors	0.14	0.35	1.92	3.21	4.92
mean number of disruptions	0.07	0.20	5.57	7.85	9.71
mean number of answers incorrect	0.0	0.0	0.92	1.92	2.40

Table 2: Newman-Keuls multiple comparison on ANOVA output for (E) errors, (D) disruptions and (C) comprehension scores

		condition 1	condition 2	condition 3	condition 4	condition 5
cond 1	E	---	SG= NS	SG= NS	SG= 38 (r= 4) p < .01	SG= 72 (r= 5) p < .01
	D		SG= NS	SG= 77 (r= 3) p < .01	SG=111 (r= 4) p < .01	SG=137 (r= 5) p < .01
	C		SG= NS	SG= 21 (r= 3) p < .01	SG= 34 (r= 4) p < .01	SG= 34 (r= 5) p < .01
cond 2	E		---	SG= NS	SG= 35 (r= 3) p < .01	SG= 69 (r= 4) p < .01
	D			SG= 74 (r= 2) p < .01	SG=108 (r= 3) p < .01	SG=134 (r= 4) p < .01
	C			SG= 14 (r= 2) p < .01	SG= 27 (r= 3) p < .01	SG= 27 (r= 4) p < .01
cond 3	E			---	SG= NS	SG= 47 (r= 3) p < .01
	D				SG= NS	SG= 60 (r= 3) p < .01
	C				SG= NS	SG= NS
cond 4	E				---	SG= 34 (r= 2) p < .01
	D					SG= NS
	C					SG= NS
cond 5	E					---
	D					
	C					

exceeded the available capacity. This overload was picked up in the comprehension score and in the performance on the secondary task (i.e. the difference in the number of disruptions reached significance level). Another result that should be noted is the non-significant difference between conditions 3 and 4 on any of the three performance measures and between conditions 4 and 5 on the disruption and comprehension scores. We conclude from these results that the level of linguistic complexity of condition 4 falls in between conditions 3 and 5 (cf. mean number of errors, disruptions and comprehension errors; table 1) and that it would suffice to test pupils on conditions 3 and 5.

The results obtained in this experiment indicate that the level of automaticity of all pupils was sufficient to perform smoothly when the difficulty level of the text was under or at their own level of performance (condition 2). This is reflected in the combined reading measure (errors, disruptions, comprehension score). However, when the complexity level was increased by making the text unfamiliar or by increasing its linguistic difficulty, the level of automaticity appeared to be too low to keep a smooth performance under dual-task conditions. Signifcantly more switches from the autonomic to the controlled mode were noted in condition 3. These switches were picked up in the disruption score and the comprehension score but not in the error score.

Close analysis of the error data and the disruption scores give some indication as to what constitutes a switch-prone element. These data data are relevant to the question of the source of linguistic complexity. The errors noted in conditions 2, 3, 4 and 5 can be subdivided into 3 main categories: (1) type 1 errors: errors of mispronunciation, adding or skipping letters and words, changing wordorder. These noted changes do not bring the text nearer to colloquial language (used by the peer group) and do not change the meaning of the text, (2) type 2 errors: errors that bring the text nearer to the pupil's own language but do not change the meaning of the text and (3) type 3 errors: errors that change the meaning of the text.

There is a tendency for type 3 errors to increase in conditions 4 and 5. Errors in conditions 2 and 3 are mostly of the type 1 category. Retrospective reports of pupils immediately after the reading performance indicate that disruptions in conditions 3, 4 and 5 mainly concern avoidance of type 2 errors ('I thought what a strange name', 'what an unusual expression', 'This person speaks funny') or type 3 errors. These data will be used to construct new reading texts for further experimentation.

REFERENCES

Boekaerts, M., Some remarks regarding the dual-task performance paradigm as a means of studying capacity-sharing interdependence in reading research (in press).

Clark, H.H., and Clark, E.V., 1977, Psychology and language: An introduction to psycholinguistics. Harcourt, Brace and Javanovich.

Coltheart, M., 1977, Critical notice on E.J. Gibson and H. Levin , eds., The psychology of reading. Quarterly Journal of Experimental Psychology, 29, 157-167.

Gibson, E.J. and Levin, H., 1975, The psychology of reading. M.I.T. press.

Goodman, K.S., 1970, Behind the eye: What happens in reading. In K.S. Goodman and O.S. Niles, eds., Reading and programme, NCTE publications.

Gough, P.B., 1972, One second of reading. In J.F. Kavanagh and I.G. Mattingly, eds., Language by Ear and Eye. Mass.: M.I.T. Press.

Guthrie, J.T., 1973, Models of reading and reading disability. Journal of Educational Psychology, 65, 9-18.

Guthrie, J.T., 1976, Aspects of reading acquisition. Baltimore, MD.: John Hopkins University Press.

Keele, S.W., 1968, Movement control in skilled motor performance. Psychological Bulletin, 70, 387-403.

LaBerge, D. and Samuels, S.J., 1974, Toward a theory of automatic information processing in reading. Cognitive Psychology, 6, 293-323.

Levelt, W.J.M., 1978, A survey of studies in sentence perception. In W.J.M. Levelt and G.B. Flores d'Arcais, eds., Studies in the perception of language. New York: Wiley.

Marslen-Wilson, W.D., 1975, Sentence perception as an interactive parallel process. Science, 189, 226-228.

Mitchell, D.C., 1982, The process of reading. New York: Wiley.

Norman, D.A. and Bobrow, D.G., 1975, On data-limited and resource-limited processes. Cognitive Psychology, 7, 44-64.

Perfetti, C.A., Goldman, S.R. and Hogaboam, T.W., 1979, Reading skill and the identification of words in discourse context. Memory and Cognition, 7, 273-282.

Rumelhart, D.E., 1977, Introduction to human information processing. New York, Wiley.

Samuels, S.J., 1970, Effects of pictures on learning to read, comprehension and attitudes. Review of Educational Research, 40, 397-407.

Shiffrin, R.M. and Schneider, W., 1977, Controlled and automatic human information processing. Psychological Review, 84, 127-190.

Singer, H., and Ruddell, R.B., 1976, Theoretical models and processes of reading. Publication of the International Reading Association.

Stanovich, K.E., 1980, Towards an interactive-compensatory model of individual differences in the development of reading fluency. Reading Research Quarterly, 16, 32-71.

Tanenhaus, M.K., 1978, Linguistic context and sentence perception (Ph. D. Dissertation). Columbia University.

TIME AND TENSE IN AN ENGLISH PEDAGOGICAL GRAMMAR

L.K. Engels, B. Van Beckhoven, T. De Bisschop, and M. Goethals

Section of Applied Linguistics
K.U. Leuven

'PICTURA EST LAICORUM LITERATURA'
Umberto Eco

According to B. Spolsky there are three kinds of pedagogical grammar in foreign-language learning and teaching; there should be a grammar for the teacher, another one for the curriculum writer and, last but not least, a different one for the students (Spolsky, 1978). This last venture on grammar should take into account not only WHAT will be taught, but especially HOW it will be presented, taught and learned. It is our intention to focus on a presentation of foreign-language grammar that will facilitate learning. For this purpose we have consulted recent developments in the psychology of learning (Paivio, 1971), namely the role of mediation in verbal learning, especially the experiments with imagery in verbal responses. We have also learned a lot from the science of semiotics, whose latest attempts to systematize the signs a human being uses (Sebeok, 1973, 1976) taught us a clear distinction between symbols ans icons. Most signs used in our approach to describe time and tense in English, are, accordingly, icons, because they can visualize even the abstract contents of a notion or a grammar rule in a simulating way. Seeing the sign in later learning stages triggers the notion or rule back into the memory. One can invent several kinds of exercises to be done by means of these mediating signs.

Time can be represented by means of an infinite time line.

Figure 1.

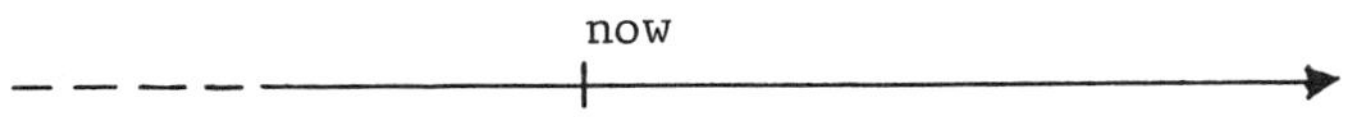

Figure 2.

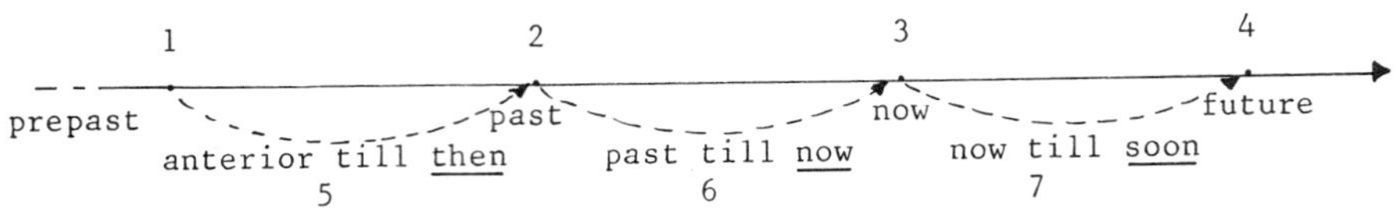

Figure 3.

To indicate the present moment, which is to be understood as the moment of communicating or speaking, we bar the time line by means of a cross section (figure 2.). Any past moment will be situated to the left of the present moment (1 and 2), any future moment (4) to the right of the NOW (3) (Figure 3.). The resulting durations (in between these fixed moments in time) can also be indicated on the time line: anterior till then (5), past till now (6) and now till soon (7).

Let's read the following text and add to each underlined verbal form the number it expresses on the time line:

"The first stage of the overseas expansion of Europe is regarded ( ) as beginning with the capture of Ceuta by the Portuguese in 1415 (on the African coast of Gilbratar). The Spaniards and Portuguese had ( ) their precursors in the conquest of the Atlantic Ocean, but the efforts of these adventurers had not changed ( ) the course of history. Vikings had voyaged ( ) to North America in the early Middle Ages. But the last of their settlements on Greenland succumbed ( ) to the attacks of the Eskimo before the end of the XVth century. Since then the whole world has been discovered ( ), occupied, colonized and decolonized. We are now looking ( ) at the planets and solar systems around us. The moon has already been visited ( ) by human beings. Will the planets and solar systems ever be reached ( )? Will mankind be imprisoned ( ) in its solar system, on this same earth?"

Feedback:

is regarded (3) - had (2) - had not changed (5) - had voyaged (1) - succumbed (2) - has been discovered (6) - are now looking (3) - has already been visited (6) - will ever be reached (4) - will be imprisoned (4).

The events or facts, actions or states expressed in this text are normally situated in a limited time frame (they have a certain

Figure 4.

duration), which is not the whole time line, but only a part of it. This observation applies, for example, to

had their precursors
had not changed
are now looking

We can use a circle in order to separate the time line into individual limited time periods (figure 4.). In this way we can focus on fragments of the time line. The circle, as it were, encompasses a tiny era of time.

Taking the moment of communication as the "now" moment, and indicating it by means of a cross section on the time line, we can represent a limited period of time at a present, past or future moment if we situate the circle either on top of, to the left of, or to the right of the present-moment bar (we needn't indicate the time line, it is always presupposed).

- Φ = present limited duration in relation to a present moment in time "the overseas expansion of Europe is regarded..."
- O| = past limited duration in relation to a present moment in time "the Spaniards...had their precursors..."
- |O = future limited time period in relation to a present moment in time "Will mankind be imprisoned?"
- O| = past-duration-till-now in relation to a present moment in time "The whole world has been discovered"
- O⁝| = anterior limited duration in relation to a past moment in time "The Vikings had voyaged..."
- O⁝| = anterior-duration-till-then in relation to a past moment in time "The efforts had not changed the course of history".
- |O⁝ = duration now-till-soon according to a future moment in time "Will the nearest star constellation have been reached within the short period of man's life?"

We used the dotted line for a past or future cross-sectional moment, when this is necessary to express anteriority to a past, or perfectivity to a future. In the other cases we do not want a second point of reference.

FACTUAL AND PROGRESSIVE ASPECTS IN RELATION TO TIME

Till now we have only spoken about limited time and moments in time. Next to these two features, the English tense system also explicitly expresses aspects by means of different verb forms. The speaker can choose the focus of his utterance, highlighting a particular attitude towards the sequence of time: he can, for instance, say either:

"My back is aching", stressing the progressivity (i.e. the

going on) of the pain in a limited duration, using an expanded verb form,

or

"My back aches", expressing the factual occurence of the pain in a limited duration, not focusing on the progressivity and using a simple verb form.

This distinction, too, can be expressed in a visible way:

⊙ factual occurence of an action, event or state, expressed by a dot • in the middle of the limited-time icon.

⊖ progressive occurence of an action, event or state, expressed as going on by means of a thick ▬ line in the circle of the limited time space.

In many example the choice of the aspects (factual versus progressive) is free; it is a matter of focus.

By means of the cross section for the present moment we can situate the factual or progressive action, event or state at a particular moment in time: present, past or future.

When the focus is on the factual aspect, we can have

- in the present ⊙ "At present we study at a university."
- in the past ⊙| "A long time ago I studied in London."
- in the future |⊙ "... and I shall have some peace there." (W.B. Yeats)
- anterior to the past ⊙¦| "The Vikings had discovered America long before Columbus."
- in the past-till-now ⊙| "We have worked hard lately."
- in the past-till-then ⊙¦ | "My friend had lived as a prisoner in a Japanese Camp till the war ended."
- now-till-soon |⊙¦ "I shall have finished this work before this term ends."

With the focus on the progessive aspect

- going on in limited time now ⊖ "I'm working hard these days."
- going on in limited time past ⊖| "I was working hard last night."
- going on in limited time future |⊖ "I shall be working for the BBC soon.
- going on in limited time anterior to the past ⊖¦| "My father had been travelling in Egypt a long time before the first World War."
- going on in limited time-up-to-the-present ⊖| "I have been working hard lately."
- going on in limited time up-to-then ⊖¦ | "I had been working for hours, when the light went out."
- going on in limited time up-to-the-future |⊖¦ "I shall have been working for hours when my friend comes back."

As has already been explained, the language-user's point of view is essential in the choice of the aspects. Sometimes more delicate differences appear:

e.g. My watch works perfectly (the watch is reliable)
My watch is working perfectly (the watch is not reliable)
The man was drawing (he can still be saved)
The man drowned (he is dead)

I was reading a book that evening (not finished)
I read a book that evening (finished)

VOICE

We can now add VOICE (active or passive) to our insights of time
{ limited duration / moment
present, past, future, anteriority
anterior-till-then, past-till-now, now-till-soon
and of aspects
{ progressive / factual
for actions, states and events.
The active voice is used in sentences with a known actor.
John locks the door when leaving his home
The passive voice is especially used when the actor is unknown, not mentioned, or unimportant; we find it very often in scientific descriptions.
signs:  etc.
The shading bars for the passive are non iconic but only symbolic, i.e. they are arbitrarily chosen and do not simulate.

The passive can be used in all tenses, with simple or progressive aspects.
e.g. Afterwards the milk is cooled to 50 degrees, at which point 3 per cent yoghurt culture is added. This is thoroughly stirred in, after which the milk is being poured into the new 250 ml. jars, with a 55 mm. opening. The yoghurt is eaten direct from the jar.
Sometimes the actor is expressed by a by-adjunct.
... large supplies of heroin had been used by practitioners for self-administration.

The village was still held by pro-communists.

SOME FURTHER IMPLICATIONS OF THE NOTION TIME

What we have studied up to now presents the essence of the English verbal system: limited time, present/past/future time, aspects, voices. All these result in a few tenses that have to fulfil a lot more tasks than described in the former paragraphs.

Once we agree that the notion of limited time is being expressed in all the tenses and appears in all the aspects and voices, it is easier for us to detect the contexts and the situations and the kinds of verbs that do not care about limited time, but that express other implications of time, i.e.:

no time
less limited time
total abstraction from time

It will occur to you that especially the present tense accomplishes a number of different tasks, because it is best suited to abstraction.

Examples:

no duration	limited duration	less limited duration or abstraction from duration	general time
He kicks the ball... and scores !	At present we study at the K. U. Leuven.	We speak English I enjoy my food.	Two and two make four

He is jumping up and down. can be used for present/past/future factual/progressive active	We are living in Brussels. can be used for all times and tenses factual/progressive active/passive	I am beginning to understand. usually restricted to present tense active/passive factual/progressive	The universe is forever expanding. restricted to present tense active/passive factual/progressive

Some explanations (several new icons had to be introduced)

1) No duration, often expressed in quick succession of actions

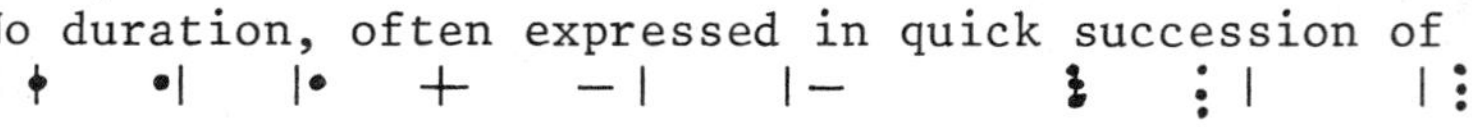

 Time after time he hits her.

2) For a lot of actions we abstract from duration and describe an ability

 He speaks German perfectly

Or we utter judgements, opinions or memories

 I think that he is ill

In these and other cases (cf. survey) we open up the circle of time-limitation so that it becomes clear that the time-limitation vanishes. Occasionally the progressive aspect () occurs (with transitional verbs).

3) When, however, the generalization is complete, we call these utterances general statements. The icon () represents a sun, radiating in all directions; it is very well suited to express "general" time.

Even the progressive aspect can occasionally occur.

The speaker can decide on the aspects, the degree of limitation or abstraction of time-duration. Compare the following examples ★
A speaker can ask a friend

"What are you doing?"
In this case he focuses on the action going in front of him.

But he can also ask

"What do you do today?"
The focus is on the factual acitivity in a limited time.

He can also express a formal introduction, in which limited time is not in focus anymore, but not quite extinguished;

"How do you do?"
We still keep the same icon, because the formula implies a state of health or general disposition, which is limited by definition.

Another question concerns the habitual activities,

"What do you do on week-ends?"
We keep the limited-time icon, because actions on week-ends still keep their nature of limited time.

The abstraction from limited time-period starts when we focus on abilities, hobbies etc., which by definition lose that time-limitation.

"What do you do (for a living)?"
"Do you fish?" (hobby)
For the first time the icon changes from time-limitation towards less-limited time, opening up the circle. Other examples:

"Do you speak English?"
"Peter works in a factory."

For more "permanent" situations we would use this icon as well,

"The picture of my parents hangs in my bedroom."

The traditionally termed "general statement" is only reached when the utterance expresses "general time", i.e. it applies to past, present and future. It is only used for eternal truths, proverbs, scientific statements etc. (more details will be supplied in the survey). The first sentence of the sample text on page 2 could be interpreted in this way; it is a passive sentence and a scientific statement.

"The first stage of overseas expansion of Europe is regarded as beginning..."

We think that the concept of a gradual movement from limited time towards abstraction of limited time, finally resulting in a total absence of time-limitation is necessary to fully understand the uses of Present Tenses in English.

★ The examples have been partly borrowed (and amplified) from a grammar for foreign learners by R. Dirven, G. Radden and D. James.

THE ASPECT OF PERFECTIVITY ◯|

The perfective aspect expresses a past action, event, or state that has a contact with either a present, a past or a future time (some later moment than the starting point of the action, event or state), and that utters its perfectivity in a result, a completion, or simply in contact.

We can show the difference between anteriority ◯¦| in which the action, event or state does not touch the dotted line, and the prepast-till-then ◯¦ | , in which the action, event or state touches the dotted line by means of two examples from the sample text.

⊙¦| "Vikings had voyaged to North America in the early Middle Ages."
Here the Plu-perfect Tense acts in the same way as the Past Tense; it expresses something definitely anterior to the past.

⊙¦| "... but the efforts of these adventurers had not changed the course of history."
Here the Plu-perfect Tense acts in the same way as the Present Perfect Tense, but in an anterior past that touches "then" (the past time).

DEFINITE VERSUS INDEFINITE TIME

The Present Perfect Tense is often used to express indefinite past. To indicate indefinite time we have added a circle that partly overlaps the circle having contact with the present moment in order to iconically represent the indefiniteness of the time.

◯◯| The contact of the time circle with the cross section of the present moment iconically expresses a past-duration-till-now

DEFINITE time indication

⊙| "Alderman Brook became a liberal representative for Birkly Ward in 1944."
Te Simple Past often expresses this definite time.

INDEFINITE time

⊙⊙| "There have been suggestions recently of reconciliation."

We shall now give a survey of this grammatical core of time, aspect, voice and tense in English. When you have fully grasped these essential notions, we are sure you can tackle our survey of the use of TENSES.

Table 1. Notions and Signs for TIME, ASPECTS, FORMS, VOICES and TENSES of the English verbal system.

		Aspect: Factual ⊙ Simple verb forms				Aspect: Progressive ⊖ Expanded verb forms			
		TIME				TIME			
		Duration			General time	Duration			General time
Aspects	Tenses	none	limited	less limited		none	limited	less limited	
cf. survey	Present	[sign]	[sign]	[sign]	[sign]	+	[sign]	[sign]	[sign]
	Past	·|	⊙|			—|	⊖|		
	Future	|·	|⊙			|—	|⊖		
Anteriority	Pluperfect	·¦|	⊙¦|			—¦|	⊖¦|		
PERFECTIVITY past-till-now	Present Perfect		⊙|				⊖|		
prepast-till-then	Pluperfect		⊙¦|				⊖¦|		
now-till soon	Future Perfect		|⊙¦				|⊖¦		
PASSIVE VOICE		does not occur	occurs [sign]	occurs [sign]	occurs [sign]	does not occur	occurs [sign]	occurs [sign]	occurs [sign]

SURVEY OF THE ACTION-EVENT-STATE FACTUAL ASPECTS IN THE SIMPLE PRESENT TENSE ACTIVE OR PASSIVE VOICE

General mediator

1. Aspect of general time (verbs used to express something without time limits)
   - eternal truths, proverbs
     e.g. True authority comes from strength, never from pouring money down the drain.
     One man's meat is another man's poison.
   - scientific statements (mathematics, geography, etc.)
     e.g. In the active state, the separation of the will from the bodily activity is so impossible to conceive that we are barely conscious of using the will to perform actions. The whole body becomes pervaded with will, is will.

2. States in a less limited time period
   verbs expressing a state:
   - inert cognition (believe, think, hope...)
   - verbs of emotion (like, hate...)
   - having and being

   e.g. He speaks English well.
   My friends live in London.
   He has a lively mind.

3. Repeated or habitual actions (states, events and actions in limited time)
   - iteration is expressed
     e.g. I earnestly desire to do God's will in all things but I often fail.
   - iteration is presupposed
     e.g. My shirts are bought at Harrods

4. States, events and actions in present limited time
   e.g. At present my friends live in London.
   We must deal with the situation as it exists today.

5. Instantaneous actions (without duration) (mostly sequence). They are very often typical "speech-acts".
   - sports commentaries, reports
     e.g. He passes the ball to Smith and Smith scores!
   - conjurors and demonstrators
     e.g. Now I put three eggs into the bowl and add one cup of cane sugar.
   - declarations (very often in the 1st person), e.g. baptism, ship launching, judge, etc.
     e.g. I name this ship Victoria.

- iteration of instantaneous actions
  e.g. Time after time a powerful undertow sweeps him outward.

6. Historic present (the dotted line of the past is drawn; an arrow points to the present used)

- summary of book or film
  e.g. On her wedding night, Mehalah hits Rebow with a bottle. It contains vitriol and blinds him.
- artists of the past and their works
  e.g. Early on in his introduction Mr. Auden comes to the odd conclusion that "if the importance of Cavafy's poetry is his unique tone of voice, there is nothing for a critic to say."
- headlines in newspapers
  e.g. Howe rules out reflation to boost economy.
  Israelis fall prey to Beirut ambushes.

- texts under photographs
  e.g. The Prince of Wales is congratulated on the birth of his son, Prince William
  Rescue workers search for victims of a car bomb explosion in the Hamra area of West Beirut.
- historical summaries
  e.g. 1942 United States and Great Britain establish War Crimes Commission
  1945 Yalta conference
  Roosevelt dies, Truman succeeds to the presidency
  War in Europe ends
- with verbs of information followed by new information
  e.g. I hear she's a most handsome woman.
  Mother tells me you're getting married.

7. Imaginary present (the line of the present is now a dotted line)
- novelists: fictional present
  e.g. Mr. Tulkinghorn glances over his spectacles and begins again lower down. My Lady carelessly and scornfully abstracts her attention. (Dickens, Bleak House)
- stage directions in plays

  e.g. Hester stands with her hand out, facing Freddie. He gives her the letter and moves away from her. She tears it up methodically and throws the pieces into the waste-paper basket. Then she takes the bottle of whisky and goes over to a cupboard. (Rattigan, The Deep Blue Seal)
- Travelogue
  e.g. To reach Chugwell, we make our way up to the source of the river Ede, then skirt the bleak slopes of Windy Beacon.

- instruction booklets
  e.g. When there is a new piece of input data, the oldest piece of data is discarded to make room for the latest input. This replacement sheme makes the moving average a valuable tool in following trends.
  Toffee apples are made with sugar. You cook the sugar until it turns into a toffee and you put your apple on a stick and dunk it into the sugar.
- verbal directions
  e.g. First you pick up the receiver. You wait for the dialling tone. Then you dial the number. When you hear the pips you put in 5 pence.

8. Present tense with future reference (an arrow points to the dotted line expressing a future moment)
- when ADV indicates future (only for proposals which are currently true and assumed to be unchangeable)
  e.g. Former Wimbledon champion Budge Patty marries on April 5 in Switzerland.

  The train leaves at 10 p.m.
- future meaning in subclause with as, before, when, until, etc.
  e.g. As soon as the leadership gets up its courage, the communes will be abolished.
  This increase will continue, as the new colleges come to completion.

9. Emphatic present (emphasis is expressed by the double underlining of the icon)
  e.g. This is no reason for Britain to behave likewise, but it does raise the problem of what will happen if Britain remains the only uncontrolled area into which the spill-over can go.

  I don't know her real husband. And what I do know I promised faithfully I'd never tell a living soul.

Check your insight ! (Use of Simple Present, active or passive)
Decoding

1) Order  (to make for) a decrease in tiredness. We  (to have) a saying "My head will never save my feet." Time after time we 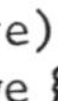 (to forget) something and have to go back upstairs or down to the shops. If we ever stopped to consider how much energy -- and time -- we  (to lose) this way in the course of a day we would be staggered. Some of it 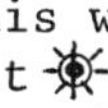 (to be) inevitable, and we  (not to want) to become too pernickety. Nevertheless, we could all probably be a little more orderly, for we so frequently just 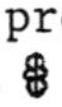 (to muddle through).

2) Tea, as a beverage, ☼ (to vary) greatly in flavour according to the water in which it ⏀ (to make). Unless you ⏀ (to be lucky) and have hit at once on a blend that ⏀ (to suit) the neighbourhood in which you ↑ (to live), you should consult a good tea merchant and get the matter satisfactorily settled once and for all. This ☼ (to apply to) both Indian and China teas. If of the latter you ↑ (to like) a "smoky" blend, mention the fact when you ⏀¦ (to ask) advice. I ↑ (not to know) how unorthodox it may be, but I ↑ (to like) to keep a piece of dried orange peel in the caddy containing China tea; purists may howl, but try it yourself.

3) "We ⧫ (to wish) to state quite clearly that it ↑ (to be) our conviction that separate development ↑ (to be) the only solution of our racial problem. We, therefore, ⧫ (to reject) integration of any form, as a solution of the problem. The agreement that has been reached ↑ (to contain) such far-reaching declarations that we cannot subscribe it. We cannot, therefore, identify ourselves with it. We further, ⧫ (to wish) to place on record our gratefulness to the Government for all the positive steps it has taken to solve the problems."

Feedback

1) makes for - have - forget - lose - is - don't want - muddle through
2) varies - is made - are lucky - suits - live - applies to - like - ask - don't know - like
3) wish - is - is - reject - contains - wish

SURVEY OF THE ACTION-EVENT-STATE FACTUAL ASPECTS IN THE SIMPLE PAST ACTIVE OR PASSIVE

general mediator ⊙ |

The action is over and time is completely past.
The speaker has a definite time in mind.
There is no strong difference between actions, events and states.

1. Any NARRATIVE, in present, past or even future

⊙ | e.g. He drew back the curtains and let the sea breeze in before he got into bed. He lay awake for a long time, thinking about the evening and the whole day.

2. Sequence of events (rapid succession is shown by dots in succession)

:| e.g. At Time Square I made a sharp left turn and went down Eighth Avenue as fast as the traffic would allow. Then I slewed into Greenwich Avenue and twisted and turned in the little side streets west of Washington Square.

3. Simultaneity in the past (two actions are shown on the same past level)

e.g. As she replaced the receiver Jenny noticed the house was completely silent again.

When I got back that window was wide open.

e.g. He was crying with frustration and anger as he whacked away, trying to chop me down like a tree.
I was trying to get through to the police when I discovered that my husband was on the line.

e.g. While the King was driving across from Balmoral to meet her, his brother, the Duke of York was performing the opening ceremony at the hospital.

I was preparing a meal for my sick wife while the children were coming home from school.

4. Referring to the PRESENT (in everyday conversation for politeness' sake)

(with a limited number of verbs of emotion or of the mind)
(= unreal past, arrow pointing to the present)

e.g. "Did you want me?" "Yes, I hoped you would give me a hand."
I thought I might go to a cinema or something.
"Are you going for one tonight?" "Yes, I thought of doing that."

5. Emphatic past (double underlining for emphasis)

e.g. I did go out for a moment. After he attacked me.
You did lose that letter, didn't you?
I went to the prison this morning to say good-bye, but she wouldn't see me. I was rather glad - she never did like good-byes.

6. Discontinued habit or state in the past: 'used to'

e.g. I used to carry their milk across a bridge in the middle of the village green. It was two feet wide with no hand rails, and I used to go across it with a black dog. The villagers didn't tell me but I know some of them used to turn out to watch me. But I never fell in.
I used to be quite handsome.

CHECK YOUR INSIGHT. DECODING

(An interview of David Owen by Robert Mckenzie, from The Listener - 25 june 1981, abridged)

R. Mck.: In your late twenties a brilliant medical career (to open) before you in research at St Thomas's Hospital on the Thames. Why •| (? you - to cross) the bridge to the House of Commons at that time?

D.O.: I ↑ (really - not to know). The first time I •| (to cross) the bridge and •| (to go to) the House of Commons, I ⊙| (to be) a Member of Parliament. I ⊚| (never - to be) there before. Very few people ⊚| (ever - to enter) Parliament knowing less. Why politics? Politics ↑ (to matter) in my family, but with a small 'p', not in a party political sense. My mother ⊙| (to be) an independent county councillor in Devon; very interested in health and mental handicap. My father ⊙| (to be) a parish councillor in the small sort of village part of Plymouth. And so politics ⊛| (to discuss). We ⊙⊙⊙| (to argue about) comprehensive education over the breakfast table. And I instinctively ⊙| (to feel) on the left, in I ↑ (to suppose), the wealthy readical middle-class tradition, but I ⊙| (not to be) very involved in party politics.

R.Mck.: When you •| (to walk out of) your medical research laboratory at St Thomas's, what ⊙| (? to be) your first feeling, going from that clinical atmosphere to the bear-pit of the House of Commons, as a backbench MP?

D.O.: It ⊙| (to be) absolute puzzlement. I couldn't find my way around the place first of all. But I ⊙| (to be very lucky), because the first day I ⊙| (to arrive) I ⊙| (to have) a message to go and see Gerry Reynolds, who ⊙| (to be) then a Minister in the Ministry of Defence, a very remarkable person, who ⊙| (to die) tragically very young. Absolutely everything that ⊙| (to be good) in the Labour Party ⊛| (to represent) by Gerry.

Feedback

R. Mck.: was opening - did you cross

D.O.: don't really know - crossed - went to - was - had never been - have ever entered - matters - was - was - was discussed - used to argue - felt - suppose - wasn't

R.Mck.: walked out of - was

D.O.: was - was very lucky - arrived - had - was - died - was good - was represented

TYPICAL USES OF THE PROGRESSIVE ASPECT

general mediator ⊖

1. Acitivity verbs

Activities and processes (cf. 2) normally have limited duration; emphasizing an action or event as going on is, therefore, quite common.

e.g. Why are you spying on me? Why can't you leave me alone? I'm not trying to decide for you whether to live or die. That choice is yours. ⊕

⊖| I wasn't asleep ... I was just reading.

Are you really working, or was that just an excuse?

2. Process Verbs

e.g. to change, to grow, to slow down, etc.

e.g. Throughout Britain indignation is growing at the cynical way in which the Government allows American bases to be set up in our land.

3. Momentary verbs (e.g. hit, jump, kick, etc.)

Momentary verbs are verbs that hardly have any duration. Using the progressive forms does not imply the notion of 'going on', but refers rather to a series of events.

e.g. He was jumping up and down.
Foster had always been hitting his servants, but with the gardener he had gone too far.

4. Transitional verbs (event-verbs like die, fall, begin, end, leave, lose etc.)

Using the progressive form indicates an approach to a transition. The action is not necessarily complete.

e.g. The sisters had to quit as their home was falling down. By May 1830 it was sadly evident that George 4 was dying.

5. Habit in Existence over a longer limited period of time (iterative)

Even though the single action or event takes up some time , the progressive form does not refer to the progressivity as such ('going on'), but it refers to a series of events.

e.g. I'm taking sewing lessons this winter.

6. Persistent activity

The events or actions are presented as going on endlessly. This is very often an exaggeration and the use of the progressive form in this case carries emotional undertones.

e.g. As a psychologist I am constantly encountering this tendency in friends.
They are like those overfussy women who are forever cleaning, cleaning, cleaning when there isn't a speck of dirt about that a man can see.

7. Planned future = anticipated happenings in future

A. Present Progressive

e.g. The Old Man's flying to Brussels on Sunday, and he wants my monthly report first thing tomorrow.

'When is Mr. Page coming home?' 'I don't know. Sometime this evening, I expect.'

B. Past Progressive = planned future in the past

e.g. Tonight, yeah, I was coming to see you about it. Now then, where's the new draft you were bringing us?

8. Progressive past for Incomplete Knowledge about recent communicative happenings

e.g. This girl you were telling me about - she is the one...? My sister was telling me about it.

9. (Rarely) General time

e.g. The universe is forever expanding.

NORMALLY FACTUAL ASPECTS; OCCASIONALLY PROGRESSIVE ASPECTS

1. Verbs of Inert Perception

Normally I can/could feel, hear, see, smell, taste
or I felt, heard, saw, smelt, tasted

e.g. I lit the cigarette and sat there and smelt the scents from the wood.
The steak is excellent, but the mushrooms don't taste like mushrooms.

Occasionally, in cases of active perception, denoting a change in the intensity of the perception

e.g. I am tasting the porridge to see if it contains enough salt.

2. Verbs of Inert Cognition (verbs of the mind)

Normally: I think, I believe, I forget, I understand
Occasionally:

e.g. You see, I was just thinking. (= ruminating)
I was just thinking, could we not go out?
I'm forgetting, you haven't met my mother.

Or for politeness' sake

e.g. We were wondering, Oliver and I, you know if you could sort of help.
I'm hoping to be a teacher when I'm demobbed.
I was hoping to see Miss Moore personally.
I think you're forgetting something.

3. Verbs of Having and Being (be, belong, consist, contain, have...)

Normally: e.g. I believe he has a fishing-rod.
The apartment belongs to George.

Occasionally, when temporary state

e.g. When I wasn't being an ineffective lay lawyer I
⊖ | was often an employment agent.
Come now, Esther, who's being devious now?
⊕ Well, the rain, now, is being warm. Soon, is being cold, like the snow.
"Freddie, please try and behave yourself." "Am I behaving badly? I'm only asking the judge here a simple question."

The expanded form of 'to have' does occur quite often, but then the verb does not mean 'to possess'. In those cases it refers to an 'experience' or 'activity'.

⊕ e.g. He's having lunch. (= is eating)
I'm having a certain amount of trouble accepting the story so far.

4. Verbs of bodily Sensation (ache, feel, hurt, itch, etc.)

Choice is free.

Φ=⊕ e.g. My knee hurts = My knee is hurting
Is someting hurting you?
| ⊙ The gag hurt her mouth.
My hair itched and I couldn't scratch it.

5. Verbs of Emotion (love, like, hate, dislike, etc.)

Seldom with progressive

e.g. I hope you are liking our hospitality

CHECK YOUR INSIGHT. DECODING

1) My door •| (to open) and a child of about fifteen •| (to put) her head round it. She ⊙ | (to look at) me for a second, wide-eyed, and then ⊙ | (to ask), " ⊕ (? I - to interrupt) you?" I ⊙ | (to assure) her that her visit was welcome, and, encouraged, she ⊙ | (to add), " ⊕ (? we - to make) too much noise?" I ⊙ | (to thank) her for her thoughtfulness and ⊙ | (to explain) that, since this was my son's party I did not feel entitled to complain. She then ⊙ | (to ask) me why I didn't come and join the party.

2) Hounds ⊕ (to scud) over the grass like a covey of grouse before the wind. You've got away on terms and the old horse ⊕ (to pull) a double handful, you Φ (to give) him his head and Φ (to let) him stride on. What else would you do when hounds ⊕ (to run)? It ↑ (to be) either go on or go home. The ground Φ (to squelch) under foot, but he can go through the dirt all day - and what a feel he Φ̲ (to give) you !

3) One of the little trials that a man must learn to bear when he

(to admit) the telephone to his home (to be) that when he (to hurry) to its side to answer a call, it will sometimes stop ringing before he (to get) there. He (to divide) the dahlias at the bottom of the garden, or (to hang) a critical bit of wallpaper in the spare bedroom, when the persisting summons (to penetrate to) his dream world.

Feedback

1) opened - put - looked at - asked - am I interrupting - assured - added - are we making - thank - explained - asked
2) are scudding - are pulling - give - let - are running - is - squelches - does give
3) admits - is - hurries - gets - is dividing - is hanging - penetrates to

THE PERFECTIVE ASPECT IN THE PRESENT PERFECT SIMPLE (FACTUAL)

general mediator

(remember: a definite past time is expressed by a Simple Past or a Past Progressive)

1. General Indefinite Past Time (two interlaced circles simulate the indefiniteness)

- with event, action and state verbs
- number of events/actions or time is unspecified
- ADV: always, never, ever, before

eg. Have you ever been to Wimbledon?
I doubt whether the overall width has ever been properly noted down.

2. Recent Indefinite Past Time (only with event/action verbs)

- ADV: just, already, recently, yet

e.g. This money - have you cashed it already?
I've just made myself some coffee to try and keep awake.
He's just gone into the garden.

3. State-up-to-the-present (contact with the present = simulated by the circle that touches the present-moment line)

- with state verbs
- ADV: since, for

e.g. All this has been out of my hands for months.
They have known each other most of their lives.
She's been very frail ever since

4. Habit-in-a-period-leading-up-to-the-present

- = repeated events/actions

- ADV: always, for, every month, etc.
  e.g. It's the one thing I've tried to avoid all my life, and yet it always seems to be happening to me.
  I've often told the children to say I'm not in when I am;
  this happens regularly with the milkman.

5. Resultative Past

- the result of a past action/event is still operative at the present moment (two arrows touching the present-moment line represent the result)
  e.g. You say you drew an extra twenty pounds a week for a whole year? Let's see your bank statement. (turns back the page) Your bank balance has dropped by over a thousand in the year.
  Mr. Dickins' methods have made Hounslow one of the most succesful teams in the country.

6. Future Reference in Subclause (the arrow points to a future moment)

e.g. I refuse to discuss the price until you've had at least three brandies.

As soon as you've delivered the goods I shall post you the cloakroom ticket and the key of the attaché-case.

Note: adverbials and tense.

a) used with past tense:
a week ago, earlier this year, last Monday, the other day, yesterday evening, just now, a moment ago, a second ago, at four o'clock in the morning on Tuesday, then, soon, next, after breakfast....

b) used with present perfect tense:
so far, up to now, since, lately, latterly, hitherto, for the present, for now, for the time being ...

c) with past or present perfect:(± interchangeable)
today, this month, this year, this century, this morning,
tonight, this March, recently, always, ever, never...

d) with past or present perfect with a clear difference in meaning:
e.g. You can tell Elizabeth that decision has now been countermanded on higher authority.

The creature had tired of playing with the raft, and now disappeared.

Look how high the hedge has already grown (as early as now)

The tank was already abandoned when I found it (as early as then)

THE PERFECTIVE ASPECT IN THE PRESENT PERFECT PROGRESSIVE

general mediator

1. Temporary Situation Leading up to the Present: Potential Incompleteness

e.g. They have been living here since their marriage.
= temporary
= they have not been married very long
They have lived here since their marriage.
= permanent time scale

e.g. It seems I've been waiting for so long.
Well, I've waited here for 2 hours; that's it. I'm off!

e.g. Someone has been drinking my beer. (=some of it is left)
Someone has drunk my beer. (= it is all gone)

In many contexts the choice simple/progressive is free.

2. Resultative Aspect

e.g. Since my marriage broke up his attentions have been slackening.
The boys have been fighting again.

3. Temporary Habit up to the Present

e.g. I followed you home that night. And I've been following you ever since.
For a whole year I've been cashing an extra twenty pounds a week.
We've been moving from place to place since they pulled down our house.

4. Ironical use

e.g. She's been leaving her husband for as long as I can remember but she never gets around to it.

5. Present perfect progressive passive

Future development of English???
e.g. The organization has been being run by volunteers.

THE PAST PERFECT, FACTUAL AND PROGRESSIVE ASPECTS

Note: The pluperfect does not merely add the perfective aspect to a past tense meaning, its meaning can be equivalent to both and present perfect.

Equivalent to Present Perfect

general mediators

1. Indefinite Anteriority

e.g. Had you ever seen him before?
I'd just spoken to my husband on the phone.

2. State-up-to-then

e.g. Then, after we had been in the desert for some weeks, one night he tried ... but I would not ... and after a short time he gave up.

3. Habit-up-to-then / Action going up to then

e.g. I was still the rector's son who, from the cradle, had been preached the abomination of carnal love.
I had worked my way up over several years.
He had been teaching in that school for 15 years.

4. Resultative-up-to-then

e.g. Filicity had reached the pinnacle of her career at last.
I got in touch with an old Oxford chum of mine who had attained distinguished rank in the police force.
Nancy's screams had not been heard because of the thick wallpaper plastered over the boards of the attic.

5. Equivalent to Present Perfect with Future Reference

e.g. It was offered to me in strict secrecy until I'd made up my mind.

Equivalent to past

general mediators

1. Anteriority

e.g. I had met him a few days earlier.
The parcel had arrived on April 15th.

I'd been out to a cinema, and I came back here
Had you arranged anything special for dinner yesterday?

2. Past and Pluperfect Interchangeable (with "before" and "after")

e.g. Originally I {had planned / planned} to shock you by writing in French.

{⊙¦| / ⊙ |} I used to "roll in" the potatoes after they {had been planted. / were planted}

Indirect Speech
In indirect speech the pluperfect is used to replace both past and present perfect tenses when the reporting verb is past.

e.g. I told him there had been an accident
= "There has been an accident".

'You said just now his feelings for you hadn't changed.'
'They haven't, Bill.'

⊙ | Betty told herself that she had been happy and content enough before she had fallen a victim to its magic spell.
= "I was happy ...before I fell..."

Mr. Gaitskell said he had been told in Europe last weekend that we were looked upon as a liability.
= "I was told in Europe ...we are looked upon.."

Check your insight. Decoding

1) Last week ⊙ | (to mark) the end of the school year. It means relaxation after a long period of intense activity, which, for many children, ⊙| (to indicate) prospects for the future. Some ⊙| (to say) farewell to schools that ⊙| (to guide) and (⊙|) (to encourage) them, and next month they will be going on to one of the forms of secondary education now bestowed. Others ⊙| (to leave) school to make their way in a highly competitive technical and scientific world.
2) I ⊙ | (to go down to) breakfast that morning with a sort of pre-examination shakes. The hot weather, or the wine at dinner ⊙| (to give) me palpitations and I ⊙ | (to feel) suspicious. I ⊙¦| (to see) the world the night before as one gigantic romance. Yet, when I ⊙ | (to join) the others, everything, from the littered table to the look on Alison's face, ⊙ | (to seem) horribly normal, horribly mundane.
3) Nothing ⊙ (to leave) to chance that careful preparation can obviate and in some cases professional coaches of Indian origin ⊙| (to employ). Without going as far as that England ⊙| (for some

time - to have) a national team coach in S.D.Dickens whose methods ⊚ (to make) Hounslow one of the most successful club sides in the country over the last ten years. Unhappily, Dickins ⊙| (so far - not to enjoy) the financial and administrative backing that a coach must have if he is to produce results.

4) For an eternity, it ⊙| (to seem), the captain ⊙¦| (to be) at strife with the elements and at strife with men. With axes and steam horses he ⊙⊙¦| (to fight) the creeping Black Frosts which ⊚¦| (to try) to burden his ship with their deadly weight of ice. With brain and furious strength he ⊙⊙¦| (to fight) the giant seas which ⊚¦| (to lick) men away like flies and (⊚|) (to threaten) to engulf him. With cruel fists and crueller words he ⊚| (to cow) crews made mutinous by wicked overwork.
5) She •| (to yawn), •| (to slip off) her shoes and, stretching out her legs, ⊙| (to lie) full-length along the seat. By the time the train ⊙¦| (to pull into) the next station, Sheila ⊙| (to be) in a half doze.

Feedback

1) marked - has indicated - have said - have guided - (have) encouraged - have left
2) went down to - had given - had seen - joined - seemed
3) is left - have been employed - have for some time had - have made - has not so far enjoyed
4) seemed - had been - had fought - had tried - had fought - had licked - (had) threatened - had cowed
5) yawned - slipped off - lay - had pulled into - was

Expression of Future Time

general mediator |○

1) Shall/will

A. Simple Future

1. Neutral predictions

|⊙ I shall/will / you will / he/she will +Inf. or I'll / you'll / he'll/she'll + Inf.

e.g. We shall send you a form for your employer. Your claim will be dealt with by post and it |● will not usually be necessary for you to be interviewed.

2. Predictability

e.g. Even though I'm not a prophet, I would say that she will probably try again to commit suicide.

⊚ She will stay here. She'll turn in early and listen to Saturday Night Theatre in bed. She always does when I'm out.

3. Real condition

e.g. Difficulties will arise over tropical products if one or more of the less-developed countries or territories of the Commonwealth do not enter into an appropriate form of association with the Community.

If South Africa is allowed to remain, Britain's prestige in Africa and Asia will dwindle as it did after the Suez escapade.

4. Prophetic statements

e.g. The Shamir was created on the eve of the sabbath, to bring the sabbath, real peace to the whole world. Then the whole earth will be his altar.

In that age Jerusalem will indeed have Jehovah dwelling in her midst and be called a city of truth.

5. Assumption about the present

e.g. He will leave you now.
It'll be alright now.
Nothing will stop us now.

B. Future Perfect

1. Action/event/state preceding a future moment

e.g. The net cost of the service to the Exchequer will have increased over 3 years by 20 per cent.

2. Assumption about a past time period

e.g. The dog's probably gone by now but I doubt if the crap will have fled.
"It's twenty past ten."
"Never mind, the au-pair girl will have gone to bed."

C. Future Progressive

1. Prediction of an action going on in a future time period

e.g. During the training week, factories all over the country will be holding 'open days'.
Soon one person in four will be living in a postwar house.

2. Assumption about the present

e.g. You will be starving now

3. Future as A-MATTER-OF-COURSE
The predicted event will happen independently of any will or intention. This is the most objective, purest future possible (a colourless future).

e.g. Will you be bringing the car?
"Sweet dreams." "They'll be sweet all right. I shall be dreaming of you."
To some extent I shall be touching on points already made by previous speakers.

The use of the future progressive is especially common in questions, where it is a politeness device. Asking "Will you have ..." would be imposing one's own opinion or will on the other person. A "Will you be having ..."-question is polite and neutral.

e.g. Are you going to put on another play?
= a direct question about somebody's intentions
Will you be putting on ...?
= a question about future activities, events
= pure future, polite, remote
= no intention or will
Will you put on ...?
= a request, "please, do!"

4. In utterances expressing irony and disapproval

e.g. That kettle will be boiling its liver out!
You'll be sleeping your head off !
He'll be selling his wife next !

5. Future Progressive Passive
e.g. "Will you be being lunched ?"
"I shall be being lunched, actually."

D. Future Perfect Progressive

e.g. Tonight the competitors will have been driving for 24 hours.

2) Be going to

A. Expression of intention (= strong expectation)

1. Present intention
e.g. This week she's going to be in a beauty contest. She's doing it mainly for the 500 dollar scholarship for the winner.
Sheila, are you ever going to tell Tony... about us?

2. Past intention

e.g. I was determined that as soon as I could I <u>was going to go</u> somewhere she hadn't to keep me.

3. Non-fulfilment of an intention

- past

e.g. What about that little cat or dog you <u>were going to bring</u> home for Hal?
The question I <u>was going to ask</u> you is too big to be put into a single sentence.

- present perfect

e.g. He'<u>s been going to put up</u> an extra shelf in the kitchen for weeks but ...

4. Backschifting of present intention

e.g. "I always like my mustard made with sherry,' she explained.
She <u>wasn't going to risk</u> having the watered kind.

B. Expression of subjective certainty

1. Present subjective certainty

e.g. It'<u>s all going to be</u> quite civilized, I can see that. I know you'<u>re going to agree</u>.

2. Past subjective certainty

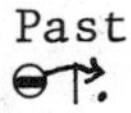

e.g. The car <u>was going to crash</u> but I brought it to safety.

3. Backshifting of subjective certainty

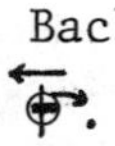

e.g. Everything was alright now, thank God. She <u>was not going to break with</u> him, after all.

3) Present Progressive with Future meaning (present plan of future happening) arrangement

e.g. I'd better phone the Grendon, and tell them we'<u>re not coming</u>.
"I think you'd better go." "<u>I'm going</u>. Don't worry."

<u>Are you doing</u> anything tomorrow night? ...
How would you like to come to a stag party?
... Some American boys have been playing tennis all over the Continent, and we'<u>re giving</u> them a sort of farewell dinner.

4) Simple Present with Future meaning

1. In dependent clauses

The present is used with future meaning in dependent clauses with: if, unless, as, as soon as, when, ...

e.g. When the phone rings you'll see the light go on under the bedroom door. When she opens it the light will stream across the room. So don't move until she picks up the phone.
We'll have a real honeymoon when I'm well.
If we drop our professional standards by turning out mass produced work the public in the end will judge us.

2. In main clauses with an adverbial of future

(same degree of certainty we normally accord to present and past events)

- fact

e.g. The muscial show makes its debut next Saturday.
At a given signal the display begins with the ignition of rockets of the cheaper range, at the north end. Then the mother patrols start a general display of smaller fireworks at the south end.
Federal Germany votes tomorrow.

- unalterable arrangement

e.g. I take him to a party tomorrow night.
He marries on April 5th.

Check your insight. ENCODING

The second case, which ( ) has also been through the courts and ( ) resulted in similar embarrassment for the police, ( ) concerns a young black man called Derek Donaldson. Donaldson ( ) sports a smart leather jacket and ( ) drives an old but impeccable BMW. One day in 1982, he ( ) was leaving a garage forecourt in Notting Hill when an unmarked car carrying plainclothes police officers ( ) came by. They ( ) ordered him to get out of his car. He ( ) was frightened that they ( ) were going tc search him and ( ) plant drugs on him, so he ( ) drove off round the corner to his mother's house. The police ( ) gave chase and ( ) summoned help. By the time Donaldson ( ) arrived at the stairs to his mother's basement flat, policemen ( ) were arriving from all directions. There ( ) was a tremendous fight, during which Mrs Donaldson's elaborate basement garden ( ) was wrecked and Donaldson himself ( ) was injured. The police ( ) found no drugs and no other evidence that Donaldson ( ) had done anything wrong. He ( ) was charged with assault on the police.

At the trial the police officers ( ) said that they ( ) had stopped Donaldson's car because it ( ) had been seen nearby in a well-known drug-dealing area -- in fact, he ( ) lives there -- and that 'it ( ) was just a routine stop and search'. Again, this either ( ) displays naivety or arrogance. There ( ) is no such thing as a routine stop and search. The judge had no hesitation in saying that the officers clearly did not have proper grounds for the stop and search, and threw out the case.

Under the provisions of the new Bill the police could almost certainly have justified their actions. Apart from the fact that the stop and search powers will now cover the whole country, instead of just selected conurbations, the powers are being widened to include searches for 'going equipped for burglary'.

There can be no doubt that the police are going to have to treat these new provisions with care. If they are seen to be stopping and searching on the basis of these trivial items, if they are seen to be taking advantage of the broadly based nature of the new law, there will be violent protest from the areas where young people and blacks congregate. The same is true of the detention provisions and the new wider powers of search of premises. If the police, as their critics allege, use the 96-hours maximum regularly, if they invade people's homes on flimsy pretexts, or simply on inaccurate information, then they will bring themselves and the law into disrepute. If, on the other hand, as Lord Denning insists, the police behave honestely and with concern for the rights of the ordinary individual, then the safeguard that the Bill provides will work well and protect society.

Feedback

⊕| (also - to be through) - (⊕|) (to result) - ↑ (to concern) - ⊕ (to sport) - ⊕ (to drive) - ⊗| (to leave) - ⊗| (to come by) - •| (to order) - ⊙ | (to be) - ⊕.(to search) - (⊕.) (to plant) - ⊙ | (to drive off) - •| (to give) - •| (to summon) - ⊗| (to arrive at) - ⊗| (to arrive) - ⊙ | (to be) - ⊗| (to wreck) - ⊛ | (to injure) - ⊙ | (to find) - ⊙¦| (to do) - ⊛ | (to charge)

⊙ | (to say) - •| (to stop) - ▬¦| (to see) - ↑ (to live) - ⊙ | (to be) - ↑ (to display) - ↑ (to be) - •| (to have) - ↑ (not to have) - •| (to throw out)

⊙ (now - to cover) - ⊕ (to widen)

⊕ (to have to) - ⊗ (to see) - ⊗ (to be) - ⊗ (to congregate) - ↑ (to be) - ↑ (to allege) - ⊗ (to use) - ⊗ (to invade) - ⊗ (to bring) - ↑ (to insist) - ⊗ (to behave) - ↑ (to provide) ⊗ (to work) - (⊗) (to protect)

(From the listener, 5 July 1984).

REFERENCES

Brown, J.W., 1978,On the use of visual and graphic codes in the production of verbal utterances. Montreal, AILA Congress.

Candelier, M., 1980, Les moyens graphiques dans les schémas grammaticaux: Quelques examples, in Bulletin de l' A.F.L.A. no. 6.

Dirven, R., Radden, G. and James, D. ( ) English Grammar for foreign Learners, Chapter 7: Time Relations.

Edmondson, W., House, T., Kasper, G. and McKeown, J., 1977, a pedagogic Grammar of the English Verb. In Tübinger Beiträge zur Linguistik, 95

Engels, L.K., De Bisschop, T. and Van Beckhoven, B., 1983, Developing Iconic Mediation in the Internalization Process of English Syntactic Rules, Ghent Communication and Cognition Colloquium.

Engels, L.K., De Bisschop, T., Leenders, T., Van Beckhoven, B., Goethals, M., Deboosere, T., 1983 , Grammicon I Exercises II. Introduction, Grammar and Keys, ACCO, Leuven.

Gagné, R.M., 1973, The Conditions of Learning. New York: Holt, Rinehart and Winston.

Glenn, E.S., 1973, Symbolic Function, particularly in Language, in Semiotica 8, 97-112.

Gregg, V., 1975, Human Memory, London Methuen.

Leech, G.N., 1971, Meaning and the English Verb., London, Longman.

Leech, G.N. and Svartvik, J., 1975, A Communicative Grammar of English, London, Longman.

Paivo, A., 1971, Imagery and Verbal Processes. New York, Rinehart and Winston.

Sebeok, T.A., 1976, Contributions to the doctrine of Signs, Indiana University, Bloomington & The Peter De Ridder Press, Lisse.

Sharwood, Smith, M., 1977, Aspects of Future Reference in a Pedagogical Grammar of English. Frankfurt, Lang.

Spolsky, B., 1978, The relevance of Grammar to Second Language Pedagogy, in AILA, Bulletin 2, 5-14.

Van Besien, F., Spoelders, M., 1973, Bedenkingen bij een pedagogische grammatica. Gent.

Zimmermann, G., 1977, Grammatik in Fremdsprachenunterricht. Frankfurt, Diesterweg.

The examples have been taken from authentic British English Corpuses: the Leuven Drama Corpus and the Lancaster-Oslo-Bergen (LOB).

CHOICE OF DIRECTIVES IN SPONTANEOUS FAMILY INTERACTION

Erica Huls

Department of Social Psychology

KUN, 6500 HE Nijmegen

1. THE CONTEXT OF THE STUDY

In discussions about the factors that might influence social inequality in school and society, it is suggested that a central role is played by the formulations through which behavior is regulated in families of different socio-economic background. In the late sixties and the early seventies, the British educational sociologist Basil Bernstein devoted a lot of his theoretical writings to this idea. His contribution to the discussion however has been rather fruitless. A lack of clarity in his writings and the unelaborated character of his ideas in the first phase were, among other things, important factors that gave rise to a polemic discussion (the difference-deficit controversy, involving e.g. Labov (1972a), Dittmar (1976), Bereiter and Engelmann (1967) and Bernstein (1973), which was hindered by mutual misunderstandings. On this point I won't go into finer details, nor try to say the final word. The only thing that really matters is the problem which Bernstein raised. His 'theory' focusses on differences in the 'communicative capacities' of children, which he tried to explain in terms of a total theory of social structure. The value of this part of his work is acknowledged by others, e.g. as early as 1972 by Hymes (1972a), and in the eighties by Hudson (1980). It is a pity that this central idea has until now elicited so little empirical research. The question which immediately comes to mind, then, is: why hasn't there been more research on this topic?

It is not possible to give the answer to this question without signalling some developments in linguistics. In what is traditionally the mainstream of thought in linguistics, language is seen as a rule-governed system, only marginally influenced by social variables. Research interests have focussed on grammar and phonology. Recently, however, a contrasting stream of thought can be distinguished (Ervin-Tripp & Mitchell-Kernan, 1977). The basis of this new line in research interests can be found in the ethnograpgy of speaking (Hymes, 1962, 1964; Gumperz and Hymes, 1964, 1972), sociolinguistics (Labov, 1972b) and conversational analysis (Schegloff, 1968; Sudnow, 1972; Turner, 1974). Some basic themes are (according to Ervin-Tripp & Mitchell-Kernan, 1977):

1 natural conversation is the data source;
2 sentences are not the highest level of analysis;
3 social context is relevant to linguistic rules;
4 variability is a component of linguistic rules;
5 language functions are diverse.

The notion 'communicative competence' is central to what can be called an interactional view on language phonomena and research. Linguistic competence (Chomsky, 1965) is only a part of what is needed for participating in language-use situations. Participants in interaction not only have to know what's formally possible, but also what is 'feasible', 'appropriate', and 'performed'. These four forms of knowledge are summarized by Hymes (1972b) as 'communicative competence'. Hymes (1967: 13), the founder of the so-called 'ethnography of speaking', argues for "a study of speaking that seeks to determine the native system and theory of speaking; whose aim is to describe the communicative competence that enables a member of the community to know when to speak and when to remain silent, which code to use, when, where, and to whom, etc. In considering what form sociolinguistic description might take, ... one needs to show sociologists, linguists, ethnographers and others a way to see data as the interaction of language and social setting".

Interactional sociolinguistics has only recently developed so far that its insights can be used for empirically investigating the problem which Bernstein stated. Bernstein was aware of the insufficient linguistic and empirical basis of his writings. He rejected Chomsky's theory of language, because it didn't conceptualize language in relation to social interaction. He was looking for an alternative theory of language, which focussed on language functions; but this theory wasn't yet available in more than a fragmentary form. This is an important factor in understanding why the linguistic aspect of the reproduction of social inequality has hardly been worked out empirically.

The present research should be seen as a methodological exercise in rendering the questions raised by Bernstein (1973), concerning the connection between social relations and linguistic forms, accessible to empirical research based upon concepts from pragmatics and conversational analysis.

For this paper I have selected the topic of the choice of directives. Families are supposed to vary in this respect. Terms to describe the difference are: authoritarian vs. 'laissez-faire', position-oriented vs. person-oriented, direct vs. indirect, imperative vs. based on appeals, normative vs. rational, and explixit vs. implicit. The first term of each dichotomy is supposed to be associated with (although not exclusively for) a lower social class, the second with a higher.

The hypothesis that families of different social class vary in the formulation of directives, with families in a higher class using more person-oriented, indirect, and explicit forms, and families in a lower class using more position-oriented, direct and implicit forms, has been tested among others, by Hess and Shipman (1967) and Turner (1973). Their results were in line with their expectations.

These studies met with severe criticism. Data were collected in unnatural situations, and this was evaluated as a shortcoming because of the context-sensitivity of linguistic data (see also under 2). The analysis of the data has been criticised from a pragmalinguistic viewpoint. Notions such as 'imperative', 'command' and 'directness' are only globally and intuitively defined. This procedure in itself is problematic. When used in the context of a normatively charged object of research, such as class-specific control strategies, it can lead to further, unreliable pronouncements. Moreover, in their method of analysis no allowance was made for the possibility that language utterances can fulfil more than one function at the same time.

Recent developments in pragmatics appear to contain points of departure for a finer and more consistent approach to the problem. In particular, the work of Searle (1975, 1976), Labov and Fanshell (1977) and Ervin-Tripp (1976) on the formulation of directives seems useful for our problem. In the present investigation an attempt has been made to apply insights from pragmatics to an extensive body of spontaneous and natural language samples from two families of different social classes.

2. DATA COLLECTION

Data were collected by participant observation in two families with children from five to six years of age.

Various considerations were taken into account in the choice of participant observation as in the method of data collection. First of all, linguistic analysis is such that it cannot be carried out exclusively on the basis of audio-sound recordings: the type of variable (pragmatic and conversational) demands more than a superficial background knowledge of the interaction. Secondly, the object of the research consists not of individual children or mothers and children, but of the social network of the family. Participant observation provides a suitable method for gaining access to complex social groups (cf. the contributions in Brunt 1977). The third and decisive reason is related to the methodological aspects of the language and social inequality discussion. In much of the research which appears to substantiate the deficit hypothesis (i.e. children of lower socio-economic background lack certain abilities, capacities, etc.), the data are collected under formal and artificial conditions (e.g. Hess and Shipman, 1965, 1967; Bernstein, 1960; Templin, 1957; etc.). This method has come under fire. It would appear that in the interpretation of results insufficient attention has been paid to the possibility that formal research conditions have varying effects on children, depending on the social environment from which they come. In particular the possibility that children from a lower social milieu are put at a disadvantage by the research situation seems to have been overlooked. The conclusions derived from formal situations have been unjustifiably generalized to behavior in natural situations. Labov makes it clear that this generalization cannot be made - at least for linguistic data: speech appears to change drastically under the influence of observation. For linguistic research this points to a difficulty which Labov (1972b: 209) summarizes as the 'observer's paradox': 'The aim of linguistic research in the community must be to find out how people talk when they are not being systematically observed; yet we can only obtain these data by systematic observation'. Because participant observation interferes relatively little with the spontaneous and natural character of interaction (Labov, 1972b), it was chosen for the present research.

Via a Family Care organisation contact was sought with families which needed help and met the criterial demands posed by the investigation. In the role of a household helper the investigator was able to enter the family as a participant observer (naturally with the consent of those involved). In this way the natural character of the family interaction was hardly, if at all, disturbed.

The observation took place daily, between 3:30 and 7:30 p.m. for a period of seven weeks. Following a period of mutual habituation between researcher and family, recordings were made of verbal interaction.

In the manner described above data were collected within two families having widely divergent socio-economic status: the family of a janitor and the family of a factory director.

In the janitor's family neither parent had completed a course of education subsequent to elementary school. Before her marriage the mother had first been employed as an unskilled factory laborer and later as a household helper. The family includes five children, all girls. The eldest, Mirjam, is eleven years of age. She attends a boarding-school. The next to eldest child is Elise, who is 9. Then comes Mieke,the six-year-old who is the focal point of the research. In addition there are two 'little mollies', as their mother calls them: Ingrid who is about one year and baby Elsje.

In the family of the factory director both parents have completed a course of higher education. The father has studied law and the mother is a kindergarten teacher. The family includes three children: Leonard, the eldest, is nine; Wientje, 'the object of the research' (as her mother puts it) is five; Otto is two.

The recordings were made with a portable cassette recorder (Philips N2223) which was set up in the living room as inconspicuously as possible. By way of trial, supplementary recordings were made in the lower class family with mobile recording equipment (the Sennheiser Mikroport).

From the recorded material a selection of two times eight hours was transcribed. One recording hour took about 60 hours to transcribe. On the basis of these data the linguistic methods of analysis were developed and carried out.

3. DATA ANALYSIS

The notion 'directive' forms the point of departure for the analysis. A 'directive' is defined as an effort to steer the nonverbal behaviour of the addressee (cf. Ervin-Tripp, 1976). This effort may incorporate various degrees of power (from strong to weak) and may vary, for example, from invite to suggest, urge, dissuade, order, or forbid.

In accordance with pragmalinguistic research by Searle (1975), Ervin-Tripp (1976) and Garvey,(1975) a typology of directives has been developed (see table 1). The categorizational principle behind the typology can be summarized as follows: what is the relation between the actual chosen formulation, the illocutionary force of the utterance (its function as e.g. a directive, representative, or expressive speech act), and its propositional content (an addressee performs an act). This relation can vary in directness. Although I cannot treat the whole table 1, I will treat three examples as an illustration.

With the imperative Just run upstairs (code 1 in table 1) the illocutionary force as well as the propositional content of the directive is formulated, so this type is relatively direct. With the formulation, Do you want to get a serving spoon for me?

Table 1. Overview of directive types and their interpretation.

code	Type + example of a formulation	Major type 1	Major type 2	Explicitness	Strategy
1	Imperative Just run upstairs	direct	direct	yes	none
2	Explicit performative I advise everyone not to lay one single finger on it	direct	direct	yes	none
3	Ellipsis A Plate	ellipsis	direct	yes	none
4	'You can' You can play outside once in a while, can't you?	matrix of felicity	embedded	yes	personal (hearer)
5	'I want you to' I want you to play in the bath for a while after supper	matrix of felicity	embedded	yes	personal (speaker)
6	'You do' And Wientje, you come with me to get some bread	matrix of felicity	embedded	yes	none
7	'You must' And your mother must go on sitting here	other matrix	embedded	yes	rights, duties, sanctions
8	'You may' Yes, but you may not go on walking around naked like that	other matrix	embedded	yes	rights, duties, sanctions
10	'Do you want to' Yeah, look, do you want to get a serving spoon for me?	other matrix	embedded	yes	personal (hearer)
11	'You need to' No, no, then you don't need to put those trousers on anymore either	other matrix	embedded	yes	none
12	'Why' Can you tell me why you don't want to go to the same school as Walt?	other matrix	embedded	yes	personal (hearer)
13	Another matrix or two matrices simultaneously When are you going to get the chocolates?	other matrix	embedded	yes	none
14	Permission directive Mom, may I put the bikini on?	indirect	indirect	no	none
15	Question directive Do you want me to get angry?	indirect	indirect	no	rights, duties, sanctions
16	Hint We're not in the Dutch Indies here	indirect	indirect	no	none
17	Other	-	-	-	-

(code 10 in table 1), the propositional content is stated (you get a serving spoon for me), but a minunderstanding is possible concerning the directive aspect of the formulation (one can simply take it as a question about the addressee's willingness). The propositional content is in this case couched in a frame (a 'matrix' in the terminology of Garvey (1975)), which brings up the question of the adressee's willingness, so this type is of an intermediate directness. The last example, code 16 in table 1, is typically Dutch, We're not in the Dutch Indies here. The situation is that a child is sitting naked at the table and her parents want her to put on some clothes. In this formulation neither the desired act (to put on clothes), nor the agent (the child), nor the illocutionary force of the utterance (a directive) is literally formulated. So this type is relatively indirect.

The left half of table 1 contains the types, on the right half the types are regrouped under more general headings. Again I can only treat some examples. Types 1 and 2 (the imperative and the explicit performative) are both interpreted as direct. Type 5 and 10 ('I want you to' and 'Do you want to') are both interpreted as embedded (under 'major type 2'). However, there's a difference in the kind of matrix which is used (under 'major type 1'). With 'I want you to' the speaker expresses his wishes, and so he embeds the propositional content in a matrix concerning one of the so-called 'felicity conditions' (Searle, 1975) in a directive speech act. With 'Do you want to' no such condition is involved. Type 15 and 16 (the question directive and the hint) are both interpreted as indirect. Type 7 and 8 ('you must' and 'you may') are interpreted as expressing the same strategy, namely referring to 'rights, duties, and sanctions".

The last two types found in table 1, the question directive and the hint, show immense diversity in content. For these types a subdivision has been made concerning 'theme': what sort of reason or condition does the speaker supply in order to move the addressee to the desired act. The speaker can, for example, call attention to his own feelings, a sanction, or a general norm. Table 2 provides an overview of the distinctions that have been made for 'theme'. Analogous to table 1, this table also gives an overview of the interpretations.

Each directive speech act in the research sample was coded in terms of tables 1 and 2. The results of this pragmalinguistic description were quantified. A directive analysis set up in this way provides a first step towards a pragmalinguistic access to concepts from the language disadvantage discussion, e.g. 'directness', 'explicitness', and 'personal vs. positional orientation'.

A second part of the analysis is more qualitative. This part, which consists of the treatment of directive fragments,

Table 2. Overview of the themes distinguished within the directive types 'question directive' and 'hint', and their interpretation.

Code	Theme and example of a formulation	Major type 1	Major type 2	Explicitness	Strategy
1	Need or feelings of the speeker I know I was always scared to death when I saw the plumber climbing through the window	indirect	indirect	no	personal (speaker)
2	Necessity It's not at all necessary	indirect	indirect	no	general
3	(Im)possibility It's not possible, dear	indirect	indirect	no	general
4	(In)ability of the speaker I can't have you on my lap the whole time, can I?	indirect	indirect	no	personal (speaker)
5	Norm or rule It's not proper to sit naked at the table	indirect	indirect	no	general
6	Sanction Alright, then I'll walk bare-arsed	indirect	indirect	no	rights, duties, sanctions
7	Consequence The spoon is going to get in there again, isn't it?	indirect	indirect	no	none
8	Statement about behaviour Come on, you're naughty	indirect	indirect	yes	none
9	Address-term and/or attention-getter Hey, little one	indirect	indirect	yes	none
10	Other And we're not in the Dutch Indies here	indirect	indirect	no	none

serves a threefold function. First, it illustrates the numerical results. In the second place, it forms a complement to the quantifying analysis by giving account of the effect of a particular directive choice. And finally, it can give insight to the sequential organization of directives (cf. Garvey, 1975), an aspect of directivity which shows so much variation, that a quantification isn't useful. Both the effect and the sequential organization of directives are worth considering in the light of the general problem of the reproduction of social inequality (see under 1).

4. RESULTS

I won't treat results concerning 'type' and 'theme' (cf. the left half of table 1 and 2), nor go into details as regards the relationship between extralingual variables (topic, situation, etc.) and directive style. This is done elsewhere (Huls, 1982). The main point of my treatment of results is to show the possibilities and shortcomings of a quantitative and qualitative pragmalinguistic analysis of directive styles in the light of notions derived from the language and social inequality discussion. This is done by a presentation of tables relating to the right half of table 1 and 2, and a treatment of fragments.

Table 3 shows results concerning major type 1. The difference between the families is statistically significant (X^2-test, $p < .001$). In the lower class family (hereafter family Low) the style is relatively indirect; in the higher class family (hereafter family High) matrices are relatively frequent, especially matrices concerning 'felicity'.

Table 3. Structure of the directive repertoire of the families, the mothers and the pre-schoolers as regards major type 1 (percentages)

	major type 1					
	direct	ellipsis	matrix of felicity	other matrix	indirect	Total
Family Low	28.9	7.0	6.0	11.8	46.3	100%
Mother as speaker	23.3	4.5	6.2	11.3	54.8	100%
Pre-schooler as speaker	24.6	17.4	7.6	9.8	40.7	100%
Mother as addressee	21.2	13.7	9.6	16.4	19.0	100%
Pre-schooler as addressee	22.5	6.9	7.5	17.7	45.4	100%
Family High	25.5	7.2	13.4	14.8	39.2	100%
Mother as speaker	28.4	5.5	12.5	11.9	41.7	100%
Pre-schooler as speaker	25.8	9.5	7.1	26.8	30.9	100%
Mother as addressee	18.3	9.6	12.9	18.3	40.8	100%
Pre-schooler as addressee	28.9	5.2	12.3	11.3	42.4	100%

Table 4. Structure of the directive repertoire of the families, the mothers and the pre-schoolers as regards major type 2 (percentages)

	major type 2			Total
	direct	embedded	indirect	
Family Low	35.9	17.8	46.3	100%
Mother as speaker	27.7	17.4	54.8	100%
Pre-schooler as speaker	41.9	17.4	40.7	100%
Mother as addressee	34.9	26.0	39.0	100%
Pre-schooler as addressee	29.4	25.5	45.4	100%
Family High	32.7	28.1	39.2	100%
Mother as speaker	33.9	24.4	41.7	100%
Pre-schooler as speaker	35.2	33.9	30.9	100%
Mother as addressee	27.9	31.3	40.8	100%
Pre-schooler as addressee	34.0	23.6	42.4	100%

The active repertoire of the mothers is also different ($p < .001$). Mother Low scores high in indirectness. Mother High scores high under the label 'matrix of felicity'. Mother High is, compared with mother Low, also more direct. The active repertoire of the children Wientje and Mieke is also different ($p < .001$). Mieke uses relatively often an indirect or elliptical form, Wientje a so-called 'other matrix'.

There's a lot of similarity in the passive repertoire of the mothers. The children are different in this respect ($p < .01$). Wientje gets more direct forms. She also gets more matrices of felicity. Mieke gets, relatively to Wientje, more indirect forms, and especially more forms under the label 'other matrix'.

Table 4 shows results with respect to major type 2. Globally they lead to the same picture as the previous table. The families have another form of indirectness ($p < .001$). In the family Low the active person and the act aren't specified in the family High those elements are embedded in a frame or matrix.

The mothers are also different ($p < .001$). Mother Low stands out in giving 'real' indirect directives. Mother High's scores on the other two points are higher.

Mieke is either direct, or indirect. Wientje makes relatively often use of forms in between ($p < .001$).

The passive repertoire of the mothers has the same structure; that of the children too. This last point is the only respect in which results concerning major type 2 differ from major type 1.

Table 5 shows results concerning explicitness. Neither the families nor the mothers are different in this respect. The

Table 5. Structure of the directive repertoire of the families, the mothers and the pre-schoolers as regards explicitness (percentages)

	explicitness		
	explicit	implicit	Total
Family Low	67.1	32.9	100%
Mother as speaker	62.9	37.1	100%
Pre-schooler as speaker	64.3	35.7	100%
Mother as addressee	62.2	37.8	100%
Pre-schooler as addressee	64.9	35.1	100%
Mother High	66.9	32.9	100%
Mother as speaker	63.5	36.5	100%
Pre-schooler as speaker	75.6	24.4	100%
Mother as addressee	62.7	37.3	100%
Pre-schooler as addressee	64.0	36.0	100%

children don't differ as addressee, but as speaker they do ($p < .01$). Wientje is relatively explicit.

Table 6 contains results as regards strategy. The families are different ($p < .001$). In the lower class family a formulation of rights, duties and sanctions is chosen more often, in the higher class family considerations of the addressee.

The active repertoire of the mothers is different too ($p < .001$). Mother High refers relatively often to considerations of the adressee. Mother Low to rights, duties and sanctions.

Table 6. Structure of the directive repertoire of the families, the mothers and the pre-schoolers as regards strategy (percentages)

	strategy					
	personal (addressee)	personal (addressee)	general	rights duties sanctions	none	
Family Low	3.3	2.5	3.8	13.5	77.0	100%
Mother as speaker	3.6	2.6	4.9	14.2	74.7	100%
Pre-schooler as speaker	5.3	2.1	4.9	15.2	72.4	100%
Mother as addressee	5.9	4.6	2.6	15.0	71.9	100%
Pre-schooler as addressee	1.2	3.5	7.0	16.7	71.6	100%
Family High	3.7	4.7	4.0	10.7	76.9	100%
Mother as speaker	2.2	6.4	4.3	7.6	79.4	100%
Pre-schooler as speaker	7.0	4.3	3.7	16.3	68.7	100%
Mother as addressee	8.7	8.3	5.6	11.5	65.9	100%
Pre-schooler as addressee	3.7	3.9	3.7	10.5	78.3	100%

Wientje and Mieke don't differ as regards strategy. Wientje and her mother do ($p < .001$). Wientje makes, more use of her own personal considerations than her mother. Further more she frequently makes use of rights, duties and sanctions. Mieke and her mother aren't different.

The passive repertoire of the mothers doesn't differ as regards strategy. The children's does ($p < .01$). Compared with Wientje, Mieke is regulated more often with reference to rights, duties, sanctions and general rules. Wientje is regulated more often with reference to personal considerations of the speaker.

The general conclusion has to be that this pragmalinguistic analysis can bring to surface differences between families in the choice of directive forms. These differences are relevant for the problem of language and social inequality. On the dimension of 'person-oriented vs. status-oriented control' (the variable 'strategy' in this study, see table 6) the picture which appears in the literature on class-specific styles of socialization is confirmed here: in the higher class family considerations of a personal character play a more important role than in the other family. By contrast in the latter family people appeal more often to rights, duties, sanctions and general rules. As far as directness is concerned (the variables 'major type 1' and 'major type 2', see tables 3 and 4), our results suggest an instructive contradiction to the expectations expressed in the literature: people are more direct in the higher class family. The results also disagree with expectations over explicitness (see table 5), since it appears that no differences exist on this point. Some results support the picture found in the literature, others refute it. One has to conclude that the picture found in the literature is an ideal type.

The following treatment of some 'real life' examples serves the threefold function described in the paragraph on data analysis.

In fragment 1 we can find the types and themes which are typical for mother Low. It also shows some elements of the directive style of mother Low which didn't become clear in the numerical results. As regards the effectiveness of her directives: mother Low is by no means an authority in her family. She chooses formulations which look strong (see e.g. table 6), but are in effect powerless. And as regards the whole sequence of directives: she shifts the boundaries of what is acceptable to her. Initially she wanted the kitchen to be tidied up. Somewhat later she doesn't mention the tidying up anymore, but tries to stop the fighting. Finally she is satisfied if the fighting takes place out of her sight and hearing. She threatens with sanctions which she doesn't carry out. At a critical moment, as in fragment 1, mother Low's directive style is inconsequent and powerless. At less critical moments, however,

Fragment 1. An illustration of the directive style of mother Low (see also the key for the symbols used in the transcript)

Key to the symbols used in the transcription

1	[	simultaneousness on the same track
2	( )	hardly intelligible
3	(little one)	assumption concerning what was said
4	//	interrupted
5	(1.6)	a pause of 1.6 seconds
6	(<) =	a so-called short pause, i.e. a pause equal or shorter than 1.8 seconds
7	(>)	a so-called long pause, i.e. a pause longer than 1.8 seconds
8	.,?:;!	normal punctuation
9	((the bell))	a noise of relevance for the verbal interaction
10	several	several persons, or no one in particular

Background information: Mieke and Elise have been making a mess in the kitchen and are romping.

Number	Speaker	Addressee	Indirect Addressee	Text	Commentary
1	Mieke	Elise		You've been fighting too	Mieke's first move shifts a part of the guilt from herself.
2	Mother	several		Children, tidy up the kitchen now, will you?	Mother chooses an address from which defines the children in terms of their role and age. She directs with an imperative which she embroiders with a tag question.
3	Elise	?		(I've done it).	Hardly intelligible.
4	Mother	several		Is that tidying up? The whole floor is cluttered up. Stop pulling one another's hair.	In the second instance mother chooses a question directive to which the children return no answer. Next she gives an argument for her directives, a hint. Before tidying up can be started, the hair pulling has to stop.
				((noises of a romp between Mieke and Elise))	
5	Mother	several	several	They can do nothing but fight, fight, fight.	Not addressing herself specifically to the children anymore, she utters a general complaint. She says with not too many words she thinks the fighting undesirable.
6	Mieke	Elise		Right, here, I'll hit you right in the face	Mieke doesn't react to mother. She tries to boss Elise by threatening with a sanction.
7	Mirjam	Mieke		Go on then, hit her.	Mirjam encourages Mieke in this with imperatives.
8	Mother	several		No, stop it. Oh, children. Children. You don't have to break one another's noddles	Mother tries to step in with an infinitive/imperative. When the attempt fails, she turns to an address form and an exclamation. Next she chooses a not unwitty formulation: the necessity to break each other's noddles is lacking. The aimed effect isn't achieved and she utters lamentations not specifically addressed to anybody.
		several	several	Ohoh. ((whispering)) Ohoh child.	
		several		Go outside and fight. Go outside and fight. Stop pulling one another's hair.	She tries to save what can be saved: she is already satisfied if the fighting takes place out of her sight and hearing. For this purpose she chooses an imperative twice. Finally she repeats her directive from 4.
9	Mirjam	Mother		I'm not doing anything.	Mother grasps Mirjam's arm. Mirjam declares herself innocent (compare 7).
10	Mother	Mirjam		You sit quietly, too.	Mirjam is urged to sit with an infinitive/imperative.
11	Mother	several		I've already told you this morning. You were at it then too. You'll go to bed early this evening, the two of you. No watching Top of the Pops. To bed. I'm serious.	Mother reminds them of an earlier conflict which has lasted the whole morning. she turns to sanctioning. She says she will stand no nonsense.
12	Mother	Elise		Well then, tomorrow I'll have some peace from you.	Mother looks forward to tomorrow: then an important source of the misery, Elise, will be out of the hous (Elise'll go to camp tomorrow).

Fragment 2. A first illustration of the directive style of mother High (see also the key for the symbols used in the transcript)

Speaker	Addressee	Text
Mother	Wientje	Oh, Wien, do you want to do me a favour? (2.6) Yes? (1.8) Beside mommy's bed is mommy's bag. Otherwise I have to go upstairs again. And my money's there. So much [money
Wientje	Mother	[shall I go and get it?
Mother	Wientje	I'd like you to.

when there's no power conflict between children and parents, the style leads to an atmosphere of 'live and let live'.

In fragments 2 to 4 we can see some typical characteristics of mother High's directive style. Before giving a directive, mother High asks for attention and cooperation. If the child agrees, she has won half the battle. No one else in the research sample made so much use of this two-step presentations of a directive.

Not finishing a turn at talk and pausing are two other means by which mother High presents her directives in more steps. In this way she gains attention and works up to a climax. Her directives get more force by turn-taking techniques.

Fragment 4 shows mother High's style with two naughty children. She asks for attention. She tries to change her children's mind

Fragment 3. A second illustration of the directive style of mother High (see also the key for the symbols used in the transcript)

Speaker	Addressee	
Mother	several	I think Wien (>)
Leonard	Mother	Uhh?
Mother	several	I think Wien (>)
Leonard	Mother	What do you think?
Mother	several	Well, Wien is eating better and nicer at the moment (3.8)
Wientje		((laughs))
Leonard		((hums))
Mother	several	than Leonard (>)
Mother	Leonard	although you're a good three years older.

Fragment 4. A third illustration of the directive style of mother (see also the key for the symbols used in the transcript)

Background information: Wientje and Leonard are playing noisily.

Number	Speaker	Addressee	Text	Commentary
1	Mother	several	Just listen to me	Mother asks for attention.
			((Leonard and Wientje are laughing loudly))	She fails
2	Mother	several	Ho, ho.	She tries another time.
			((Leonard and Wientje roar with laughter))	She fails again.
3	Erica	Mother	What's all this about?	Erica sees the case as rather hopeless. Indirectly she lets Wientje and Leonard know that she too is not satisfied with their behaviour.
4	Mother	several	Eh, you may only look at Wicky the Viking if you get into your pyjamas now first.	Mothergives two connected directives, one of the type 'you may' and one of the type 'you do'. The first one may only be obeyed if the second one is obeyed. Both directives aim finally at restoring order.
			((Leonard and Wientje are screaming))	Leonard and Wientje don't react to mother's directives.
5	Mother	Leonard	Have you got what I said?	Mother addresses herself with a question directive to one of the two children
6	Leonard	Mother	Y-yes.	Leonard can hardly escape an affirmative anwser.
7	Mother	Wientje	And you, have you got what I said?	Next mother addresses herself to the other child.
8	Wientje	Mother	Yes.	Wientje also has to say 'yes'.
9	Mother	Wientje	Do you just want to get into your [pyjamas then?	Mother addresses herself specifically to Wientje with a question about her willingness to perform the desired act.
10	Leonard	Mother	[Yes	Leonard gives the answer, although the question was addressed to Wientje
11	Mother	Leonard	And you get into your pyjamas [when she is back	What mother wants from Leonard, she puts in the declarative mode.
12	Wientje	several	[One two, one two, one two.	Wientje goes away.
			((Leonard keeps laughing))	
13	Mother	Leonard	Leonard.	Mother calls Leonard to order with an address form.
14	Leonard	Mother	Yes. (>) Yes mommy	Leonard interprets 13 as a request for attention. He gives the floor to mother. Mother neglects this. He gives her the floor again.
15	Mother	Leonard	Can't you behave normally?	Mother directs with the type 'you can'.
16	Leonard	Mother	Yes, but Wientje is behaving so strange	In his answer Leonard shifts the guilt to Wientje.
			((Leonard gasps))	

by ordening them to get into their pyjamas and promising a reward. When it's getting critical, she addresses herself to each child separately, so that they can't neglect her, and sends them upstairs. So she prevents other problems.

The conclusion is that mother High is in command. She tactfully makes use of turn-taking techniques and stands firm.

Because of the approach in depth of this study, results cannot be generalized to social classes. However, this was not the goal. We tried to develop an interactional sociolinguistic approach to the reproduction of social inequality, using natural data. The results allow a deeper understanding of the possible variation in communication habits with which children from different social backgrounds grow up. The children in the higher class family are used to discipline; the lower class children on the contrary go their own way. Although the interaction at school wasn't actually investigated, there's some indirect and tentative evidence (e.g. Sinclair and Coulthard, 1975; Edwards, 1976), that the nursery school child from the higher class, compared with her counterpart from the lower class, is from an early age better equipped for participating in school and society. Results of other analyses not reported here (an analysis of turn-taking and an analysis of so-called 'school speech acts', Huls, 1982) support this conclusion.

REFERENCES

Bereiter, C., and Engelmann, S., 1966, Teaching disadvantaged children in the preschool, Englewood Cliffs, New Jersey: Prentice Hall.

Bernstein, B., 1960, Language and social class, British Journal of Sociology, XI, pp. 271-276.

Bernstein, B., 1973, Class, codes and control, vol. 1: theoretical studies towards a sociology of language, St. Albans Herts: Paladin.

Brunt, L, ed., 1977, Anders bekeken: wet en werkelijkheid in sociaal onderzoek, Meppel: Boom.

Chomsky, N., 1965, Aspects of the theory of syntax, Cambridge, Mass.: MIT Press.

Dittmar, N., 1976, Sociolinguistics; a critical survey of theory and application, London: Arnold.

Edwards, A.D., 1976, Language in culture and class, London: Heinmann.

Ervin-Tripp, S., 1976, Is Sybil there? The structure of some American English directives, Language in Society, 5, pp. 25-66.

Ervin-Tripp, S., and Mitchell-Kernan, C., eds., 1977, Child discourse, London: Academic Press.

Garvey, C., 1975, Requests and responses in children's speech, Journal of child language, 2, pp. 41-63.

Gumperz, J.J., and Hymes, D., eds., 1964, The ethnography of communication, American Antropologist, 66, special publication.

Gumperz, J.J., and Hymes, D., eds., 1972, Directions in sociolinguistics: the ethnography of communication, New York: Holt, Rinehart and Winston.

Hess, R.D., and Shipman, V.C., 1965, Early experience and the socialization of cognitive modes in children, Child development, 36.

Hess, R.D., and Shipman, V.C., 1967, Cognitive elements in maternal behavior, in Minnesota Symposia on Child Psychology, vol. 1, J.P. Hill, ed., pp. 57-81, Minneapolis: The University of Minnesota Press.

Hudson, R.A., 1980, Sociolinguistics, Cambridge: Cambridge University Press.

Huls, E., 1982, Taalgebruik in het gezin en sociale ongelijkheid; een interactioneel sociolinguistisch onderzoek, proefschrift Nijmegen: eigen beheer.

Hymes, D., 1962, The ethnography of speaking, in Anthropology and human behavior, Gladwin T. and Sturtevant W.C., eds., pp. 13-53, Washington D.C.: Anthropological Society of Washington.

Hymes, D., 1964, Formal discussion, in The acquisition of language, Monographs of the Society for Research in Child Development, Bellugi U. and Brown R., eds., 29, 1 pp. 107-112.

Hymes, D., 1967, Models of the interaction of language and social setting, Journal of Social Issues, 23, 2, pp. 8-28.

Hymes, D., 1972a, Introduction, in Functions of language in the classroom, Cazden C.B., John V.P. and Hymes D., eds., New York, London: Teachers College Press.

Hymes, D., 1972b, On communicative competence, in Sociolinguistics, Pride J.B. and Holmes J., eds., pp. 269-293, Harmondsworth: Penguin.

Labov, W., 1972a, Language in the inner city, Philadelphia: University of Pennsylvania PRess.

Labov, W., 1972b, Sociolinguistic Patterns, Philadelphia: University of Pennsylvania Press.

Labov, W., and Fanshel, D., 1977, Therapeutic discourse, London: Academic Press.

Schegloff, E., 1968, Sequencing in conversational openings, American Anthropologist, 70, pp. 1075-1095.

Searle, J.R., 1975, Indirect speech acts, in Syntax and semantics vol. 3: speech acts, Cole P. and Morgan J.L., eds., pp. 59-82, New York: Academic Press.

Searle, J.R., 1976, A classification of illocutionary acts, Language in Society, 5, pp. 1-23.

Sinclair, J.McH, and Coulthard, R.M., 1972, Towards an analysis of discourse; the English used by teachers and pupils, London: Oxford University Press.

Sudnow, D., ed., 1972, Studies in social interaction, New York: Free Press.

Templin, M.C., 1957, Certain language skills in children: their development and interrelationships, Minneapolis: The university of Minnesota Press.

Turner, G.J., 1973, Social class and children's language of control at age five and age seven, in Class, codes and control, vol. 2: Applied studies towards a sociology of language, Bernstein B., ed., pp. 93-119, London: Routledge and Kegan Paul.

Turner, R., ed., 1974, Ethnomethodology, Harmondsworth, England: Penguin.

REGISTER THEORY AND COMMUNICATIVE TEACHING

Anne-Marie Simon-Vandenbergen

Department of English Linguistics

State University Ghent

0. INTRODUCTORY REMARKS

In this paper the term 'register' will be used in the sense of language-variety according to use. There are two important implications: (i) A difference of register is recognized when there is both a difference of language and a difference of use or situation. If therefore on non-linguistic grounds we distinguish two situation-types but we find no linguistic differences, we have only one register (Halliday et al., 1964); (ii) Registers are not 'marginal' varieties of a language but cover between them the whole language. Every language has its own specific register-system, which covers a range of situations. It follows that the foreign learner will have to acquire a knowledge of the register system in the target language. While register awareness in the native speaker grows partly spontaneously (by acquiring the mother-tongue in context), and partly through formal education, the foreign learner obviously goes through the latter process only. This creates the need to explain register differences.

The present paper first looks at some of the problems involved in the description and the teaching of registers and at directions in which further research would be most fruitful. In the second part a 'compromise' solution is suggested to the problem of teaching registers to the advanced general purpose learner of English.

1. REGISTER IN THEORETICAL, DESCRIPTIVE AND APPLIED LINGUISTICS

1.0. Firth and the Study of Register

While the term 'register' was first used by Reid (1956) to refer to situationally-determined varieties of language, it is Firth who must be given credit for repeatedly emphasizing the importance of 'context of situation' in descriptive linguistics. In various papers (1935, 1950, 1951, 1957), he pointed out that meaning should be seen as situational relations, as 'modes of behaviour' in relation to the other elements in the context of situation. Though Firth's view of meaning has been criticized by later semanticists, it cannot be denied that his work was fruitful in at least three important ways:

(i) He may rightly be called 'the father of institutional linguistics' (Halliday et al. 1964) in that he pioneered in drawing attention to 'context of situation' as a group of related categories such as: A. The relevant features of participants (persons, personalities) - the verbal and non-verbal action of the participants; B. The relevant objects; C. The effect of the verbal action (1950). Firth admitted that this schematic construct called 'context of situation' was very rough, but emphasized that it underlined the sociological component. It provided in any case the basis on which later constructs were built.

(ii) By emphasizing the social aspect of language (cf. Firth, 1950: 'Linguistics... is mainly interested in persons and personalities as active participators in the creation and maintenance of cultural values, among which languages are its main concern') he laid the foundations of the socio-semantic direction in linguistics (see especially Halliday, 1973, 1978), which defines competence as 'what one can do' as opposed to 'what one does', where 'do' equals 'mean'. In other words, at the heart of this approach lies the notion of meaning potential as the sets of options in meaning that are available to the speaker-hearer [1].

(iii) Firth recognized the importance of social roles in foreign language teaching and emphasized the need for a communicative approach, even though he did not call it by that name [2].

It is difficult to overestimate the influence of Firth on the development of the three directions in theoretical, descriptive and applied linguistics referred to above. His views have in any case greatly contributed to a school of linguistics which, diversified as it may be, assigns primary importance to context of situation and the social aspects of language in general.

The following sections will examine somewhat further in what way these three directions have to 'co-operate' in order to achieve further results in each of them separately. It is claimed here that the theory and description of register, systemic linguistics with its functional explanation of language, and the communicative approach in language teaching look at the same phenomena from different angles, and that they are in fact inseparable.

1.1. Register Theory and Description

1.1.1. Crystal & Davy (1969) reject the term 'register' because according to them it has been used indiscriminately to refer to the language of newspaper headlines, church services, sports commentaries, popular songs, advertising and football, etc., although there are 'very great differences in the nature of the situational variables involved' (1969:61). Their main criticism is that too much theorizing was done before the necessary data had been gathered. The claim that too little detailed analysis has been done to allow us to set up registers, while the theoretically interesting question concerning which variables determine the selection of which options at different levels can only be answered on the basis of descriptive data, empirically collected.

Crystal & Davy's solution is a methodological framework in which the 'style' of a text can be described on the basis of eight dimensions of situational constraint. The advantage of this approach is that (partial) descriptions of particular texts are arrived at, from which generalizations can be made if large samples are statistically analysed. There are, however, a number of problems connected with it, too. First, the avoidance of the term 'register' does not prevent the approach from falling into the same trap: for instance Crystal & Davy do not prove that the language of 'newspaper reporting' is a variety with its own characteristics; secondly, in the practical analyses only sporadic reference is made to the dimensions of situational constraint set up in the introductory chapters. In other words, it is not shown in what way particular linguistic features are linked to particular categories; thirdly, the dimensions of 'province' and 'modality' are not clearly defined; finally, no criteria are given for deciding what features to select for counting.

Other examples of studies which have singled out one particular register for description are Leech (1966), Huddleston et al. (1968), Charrow, Crandall, Charrow (1982).

1.1.2. The logical approach. Notwithstanding Crystal & Davy's caveat, interest in registers has received new impetus in recent years. Logicians have extended the notion beyond its strictly linguistic sense. Vandamme, in a number of his publications (e.g. 1977, 1981, 1982) certainly uses the term in a wider sense than that which it traditionally has in British linguistics, including not only verbal but also non-verbal signs, actions, perceptions and emotions present in a certain action-type. In one respect Vandamme's viewpoint is related to Halliday's view of language as 'meaning potential' referred to above. One may compare in this respect Vandamme's definition of 'register' as 'a network of possibilities'. In each such register we have several possible actions and several possible perceptions and verbal expressions related to it and vice versa. Such a set of possibilities dependent on an action type we call an 'abstract register' (1977). A concrete register is seen as an actualization of an abstract register, or of a series of abstract registers. This means that one abstract register can be realized in various ways. In fact this distinction between abstract and concrete register can be 'translated' into 'descriptive linguistic' terms as follows: no text will display all the features of the register to which it belongs. In other words, one has to start from large samples of linguistic data in order to arrive at general statements on a quantitative basis. Registers are indeed 'idealizations' at a high degree of generality. It seems to me that among the differences between Vandamme's term 'register' and the sense in which it is used in this paper, the following is important in the present context: Vandamme starts from action-type rather than from language [3], which means that the principle adopted here that linguistic difference is an essential criterion for register difference is not strictly followed in his model. Connected with this is the still unsolved problem (also mentioned by Crystal & Davy) of how general or how specific a register can be, before it loses its usefulness from a descriptive point of view [4]. The point where the descriptive approach and the 'logical' one can fruitfully meet is in the break-down of a speech situation into speech events, seen as comprising one or more speech acts. (See also further on Munby's 'micro-functions', section 1.3. below).

1.1.3. The 'grid' approach Another approach starts from particular dimensions, rather than from texts selected on an intuitive basis as being representative of particular varieties (registers). An example of such a study is Ure (1971), who shows that lexical density is a function of medium and of function. Other studies based on this method are Martin (1981a, 1981b). In this approach one variable or two in combination are selected and the relative distribution of linguistic correlates is examined in a great number of texts. The advantage of this methodology is that in this way one may eventually discover the exact relationship between 'distinctive' features and contextual categories, a relationship which, as already pointed out, is still very unclear. (see e.g. White, 1974; Halliday, 1978).

The 'grid' approach, as proposed by Ure (1982) would therefore seem to be the most promising direction to take. It involves building up a corpus of texts which is then subclassified several times over to provide contrasts on all situational dimensions. Unfortunately, very little has been done as yet [5].

1.2. Register and the Socio-Semantic View of Language as a Form of Behaviour-Potential

Halliday (1973, 1978) explicitly separates a theory of registers from his concern with the more fundamental question of the relation between the functions of language and the nature of the linguistic system. However these two concerns cannot be separated, since each profits from progress in the other field. The concept of metafunction as formulated by Halliday in a number of his publications (e.g. 1970, 1973, 1978) refers to the fact that the structure of the clause reflects the three basic functions of language: the experiential (ideational) function (for the expression of 'content'), the interpersonal function (for establishing and maintaining social relations), and the textual function (enabling the speaker/writer to construct texts and the listener/reader to distinguish texts from random sets of sentences). Systems at clause rank can be divided into three groups: transitivity, mood and theme. In Halliday's model, transitivity encodes experiential meaning, mood encodes interpersonal meaning and theme encodes textual meaning. The link with register theory is clear if we consider Halliday's definition of register as 'the meaning potential that is deployed in a given social context' (quoted from Gregory & Carroll 1978:11). It is hypothesized that register categories can be related to metafunctional components in lexicogrammar. Gregory & Carroll suggest that field of discourse determines the selection of ideational meanings, functional tenor of discourse determines the selection of interpersonal meanings and mode of discourse determines the selection of options in the textual systems. Martin (1981a) points out the problems connected with assigning systems to one particular metafunctional component but concludes that Halliday's proposal 'represents the theory's most powerful insight into language as a social semiotic, giving the model explanatory power far beyond that formalised in other generative grammers'. Martin suggests that more experimentation should be done in manipulating the contextual variables of field, mode and tenor (cf. the 'grid' approach above), and seeing what parts of the grammar are affected. On the one hand further empirical research on registers should provide more evidence for the functional explanation of language within a systemic theory. On the other hand the systemic model provides a generative framework in which register variation finds a place and can be adequately handled.

1.3. Register and Communicative Language Teaching

The concept of linguistic variation has also found general acceptance in the field of applied linguistics. The idea of the teaching of L_2 as a single, homogeneous unity has been abandoned and emphasis is now placed on the 'kind of communication' the foreign learner will need to understand and produce. This view has led to the production of numerous ESP courses, although there is still a definite lack of materials for the advanced general purpose learner, due to the lack of data on registers. If progress is to be made in communicative teaching more data on register variation is necessary. The emphasis in the literature on syllabus design and in the production of course materials for ESP, especially science and technology, is due to two factors (i) it is here that the need to 'specialize' had become most obvious; (ii) it is here that the most data are available. (See e.g. Huddleston et al., 1968; White, 1974; Allen & Widdowson, 1974). The present section intends to draw attention to the need for register studies and communicative syllabus design to work closely together.

Munby (1981) (like Halliday: cf. above) explicitly separates his primary interest - learner's needs - from the study of registers (1981: 30-32). Again, it is clear that the two cannot be separated: it is impossible to arrive at any results in communicative teaching without further research into intralinguistic variation. I will now look at Munby's model in somewhat more detail.

A central notion in the model is the communicative event. After having stated the relevant input information from sources such as: purposive domain, setting, interaction, instrumentality, instructions are given to 'state the main and other communicative events that the participant is required to handle in English, and then specify for each event its activities and subject matter' (1981: 99). These activities then need to be processed into 'micro-functions' before any decisions about utterance realisations can be made. Micro-functions are for instance: persuasion, warning, invitation, advice, urging, etc. They are to be distinguished from communicative activities, which are higher order units at the behavioural level of the communicative event (p. 46). The micro-functions are thus inbetween the communicative needs and the linguistic utterance: the needs are specified in terms of units of meaning or micro-functions (sociosemantic processing) before being realized as utterances (linguistic encoding). The micro-functions are semantic sub-categories of a predominantly pragmatic and logical kind. Whereas an activity is not ready for verbal realisation, a micro-function is. Units of meaning are thus encoded, 'taking account of dialect and target level requirements, and referring back to the profile as necessary' (p. 146). It is clear, however, that this presupposes that one knows how micro-functions are realized in different registers, since the linguistic form will vary with differences in context. This entails the problem of (i) specifying the complete number of micro-functions (partial lists are insufficient), (ii) making explicit the realisation of micro-functions [7]. Further, Munby's specifications of purposive domain, interaction, instrumentality, etc. can be translated into the contextual categories of field, mode, functional and personal tenor. In this way the registers which the learner will need to master, either actively or receptively, can be specified. The linguistic realization of the different selections from the transitivity, mood and theme networks depends on the 'interaction' of selections from these four contextual categories. Some progress has been made for particular registers of which partial descriptions exist, especially the language of science, classroom interaction (Sinclair & Coulthard 1975; Coulthard & Montgomery, 1981) and doctor-patient communication (see e.g. Candlin 1976). Information on variation along the formality axis is also found in Leech & Svartvik (1975). However, much more empirically founded information is needed before intuition can make room for rules.

2. A PROVISIONAL COMPROMISE SOLUTION: READING AND LISTENING AS MACRO-SKILLS

Since the explicitation of register differences has only just started it is especially the 'general purpose learner' who is left out in the cold. An example of the type of learner meant here is the advanced learner of English training to be a teacher of English as a foreign language. Apart from the specific register he will need for classroom discourse, he needs to come as close as possible to the native speaker's intuitions about intra-linguistic variation. The problem is that he has often learned grammar and vocabulary in a social vacuum.

It is important to keep in mind however that the educated native speaker uses a small number of registers actively, but is 'passively' aware of a much larger number. This means, for instance, that he is aware of what is appropriate or 'normal' in advertising, newspaper reports, the Queen's speeches, a scientific textbook, an administrative document, etc., without being able to produce those varieties himself. When asked to produce them he may be able to give a 'stereotype', i.e. an imitation with those salient features that have risen to general consciousness. The question is in what way the foreign learner can acquire or refine those intuitions by means of formal instruction. It is suggested here that at least something can be done with a 'guided inductive approach', i.e. exposure to data with guidance [8]. Following Munby, reading and listening (the 'receptive' skills) are seen here as macro-concepts, which are broken down into micro-skills. Reading and listening thus involve the development of different skills at a number of levels: the phonetic/graphetic, phonological/graphological, grammatical, lexical, discourse, informational, rhetorical and implicational levels [9]. This approach is outlined below.

(i) the phonetic/graphetic level

Recognition of the graphetic meaning of a written text at an advanced level involves for instance familiarity with the function and variation in letter-type and lay-out (paragraph division, etc.). One may think of the distinctiveness at this level of such varieties as legal documents, popular vs serious newspapers, scientific textbooks, the 'small print' in contracts, housing advertisements, official forms, etc. Recognition of the phonetic meaning of a spoken text involves recognizing meaningful variations in voice quality (paralinguistic variation), in contexts such as advertising, airport announcements, etc.

(ii) the phonological/graphological level

Comprehension at the phonological level involves acquiring expectations with regard to the appropriateness of meaningful variations in segmental and non-segmental phonology. Examples of the former are assimilation, elision in rapid, colloquial speech, liaison, etc. (see Gimson, 1973). Examples of the latter are hesitation features typical of informal spontaneous speech, such as pauses, irregular tempo, etc. (see Abercrombie, 1963; Crystal & Davy, 1969), the prosodic features typical of radio news broadcasts, church sermons, etc. Comprehension here also implicates recognition of attitudinal meanings of intonation patterns.

The understanding of the graphological meaning of a text involves familiarity with punctuation rules and conventions, and the way in which they are used in different registers. (Compare for instance legal documents, personal letters, private diaries).

(iii) the grammatical level

Expectations should be built up with regard to the appropriateness and effect of grammatical forms, such as for instance the structure of noun phrases, the use of the passive, 'impersonal constructions' [10], left dislocation, etc.

(iv) the lexical level

At this level a distinction must be drawn between 'technical items', and 'common core' items which may characterize registers by their relative frequency. For examples of expressions which characterize informal conversation by their frequency, see Crystal & Davy (1969). Another example is the high frequency of verbs and nouns of communication in political news reporting, a study of which is a vocabulary exercise on its own [11].

The foreign learner's attention should also be drawn to differences along the formality axis. Further, registers are often characterized by specific collocations (see e.g. Crystal & Davy, 1969: 56).

(v) the discourse level

Understanding at the level of discourse involves awareness of different types of inter-sentence linkage and their appropriateness in different registers. Halliday (1980) suggests that the kind of conjunctive relations found in a text will be characteristic of the register to which it belongs. In recent years a number of studies have been undertaken in this area. (See Martin, 1981d, 1981e; Taylor, 1983). At this level more research is needed, but it is no doubt a promising area.

Studies on the structure of spoken discourse have contributed to our understanding of the nature of coherence in a number of types of verbal interaction (See Coulthard & Montgomery, 1981).

(vi) the informational level

Understanding at this level goes beyond the understanding of individual words in that it requires familiarity with particular registers, not only with regard to field, but also with regard to tenor. Research has shown that familiarity with subject-matter certainly facilitates text comprehension (see Baten, 1981). The influence of familiarity with other contextual categories on comprehension needs further investigation but it seems again a promising research area. An example is Holland & Redish (1982) on the difficulty of understanding official forms.

(vii) the rhetorical level

Understanding the rhetorical meaning of a text involves recognition of the communicative purposes of the writer; whether the text purports to narrate, describe, instruct, persuade, amuse, etc. If one fails to recognize the intentions of the writer, nothing has been understood. (See Nash, 1980:157). At this level research into speech acts has a lot to contribute. (See Martin, 1981c). The notion of 'key' (Hymes, 1972) is also particularly relevant for understanding such intentions as irony, sarcasm, etc.

(viii) the implicational level

Understanding at this level implicates knowledge of the cultural context. This requires background information, not just on the literature of the 'target country' (for understanding literary quotations) but equally on everyday contexts: customs, political situation, school systems, television programmes, etc.

It goes without saying that there is a great amount of overlap between the different levels, but they can be seen to be what Firth (1935) called 'a serial contextualization' of facts, context within context.

3. CONCLUSION

Since Aarts (1976) concluded his paper on intralinguistic variation by saying that 'For the time being it looks as if a complete sociolinguistic grammar of English is a pipedream', enormous progress has been made in different areas. Renewed interest in the study of register variation has led to better insight into the direction which research should take, and to the beginning of an integration of register theory in a generative framework. On the other hand there are still too many gaps in the data available for register descriptions to be directly applicable in teaching, and communicative teaching courses can profit enormously from progress in register studies.

NOTES

[1] . Halliday (1973:51) professes his indebtedness to Firth: "A 'system' as the concept was developed by Firth, can be interpreted as the set of options that is specified for a given environment".

[2] . Firth (1950) points out that when he was consulted by the Air Ministry on the outbreak of war with Japan, he 'saw at once that the operating of reconnaissance and fighter aircraft by the Japanese could be studied by applying the concept of limited situational contexts of war, the operative language of which we needed to know urgently and quickly. We were not going to meet the Japanese socially, but only in such contexts of fighting as required some form of spoken Japanese'.

[3] . See e.g. Vandamme (1977)' ..... when determining the action types (these are the registers) in a certain region of human activities, we will have to make idealizations...'.

[4] . The example of the dream, discussed in Vandamme (1981) may serve as an illustration. In the account of the dream, Vandamme recognizes four registers, which rather seem to correspond to 'episodes' in the sense of Van Dijk (1982), i.e. sequences of propositions that can be subsumed under a macroproposition, hence semantic units. From a 'descriptive linguistic' point of view, there do not seem to be any noticeable switches in language corresponding to changes in contextual variables.

[5] . Linguists do not agree on the number and type of situational variables involved either. There is no need to give a survey here of the different frameworks, as this is provided by Martin (1981d), who gives an outline of register categories in the development of systemic theory (Halliday et al., 1964); Gregory, 1967 ; Ellis & Ure, 1969; Halliday, 1978; Gregory & Carroll, 1978). Martin himself decides on four categories: field, personal tenor, mode, and functional tenor.

[6] . Munby (1981) was chosen as the basis of the discussion because it is a thorough and comprehensive analysis of the communicative

needs of the foreign learner. Furthermore, it takes into account previous thinking on communicative teaching (e.g. Wilkins; Widdowson; Candlin).

7. Munby's micro-functions correspond to some of Wilkins' subcategories of communicative function or Widdowson's rhetorical acts, both of which subsume illocutionary acts (p. 46).

8. The term is borrowed from Corder (1973).

9. These levels are based on Strevens (1977:111), though certain modifications are introduced. Strevens distinguishes (for reading): iconic, linguistic, logical, rhetorical and implicational levels.

10. See e.g. Taylor (1983) for an interesting study on ways of achieving 'a feeling of objectivity' in educational English.

11. See A.M. Simon-Vandenbergen (1984).

REFERENCES

Aarts, F.G., 1976, 'The description of linguistic variation in English : from Firth till the present day'. English Studies 57.

Abercrombie, D., 1963, 'Conversation and spoken prose'. English Language Teaching, 18

Allen, J.P.B., Widdowson, H.G, 1974, 'Teaching the communicative use of English'. Iral vol. XII/1.

Baten, L., 1981, Text Comprehension: The parameters of difficulty in narrative and expository prose texts. A redefinition of readability. Unpublished dissertation. University of Illinois at Urbana.

Bazell, C.E., Catford, J.C., Halliday, M.A.K, Robins, R.H., 1966, In Memory of J.R. Firth. Longman.

Bellert, I., Weincartner, P., 1982, 'On different characteristics of scientific text as compared with everyday language texts'. In: Kittredge & Lehrberger, eds.

Candlin, C.N., 1976, 'Doctors in casualty: applying communicative competence to components of specialist course design'. Iral vol. XIV/3.

Charrow, V., Crandall, J., Charrow, R., 1982, 'Characteristics and functions of legal language'. In Kittredge & Lehrberger,eds.

Corder, P., 1973, Introducing applied linguistics. Penquin.

Corder, P., Roulet, E., eds., 1973, Theoretical linguistic models in applied linguistics. Brussels, Aimav.

Coulthard, M., Montgomery, M., eds., 1982, Studies in discourse analysis. Routledge & Kegan Paul.

Crystal, D., Davy, D., 1969, Investigating English style. Longman.

Crystal, D., 1975, The English tone of voice. Arnold.

Dieckmann, W., 1969: Sprache in der Politik. Einführung in die Pragmatik und Semantik der politischen Sprache. Carl Winter Universitätsverslag.

Dijk, T.A., van, 1982, 'Episodes as units of discourse analysis' In: Tannen, ed.

Dubin, J., 1982, 'What every EFL teacher should know about reading'. Forum XX/3.
Ellis, J., 1966, 'On contextual meaning'. In: Bazell et al., eds.
Firth, J., 1935, 'The technique of semantics'. In: Papers in Linguistics 1934-1951. Oxford U.P.
Firth, J.R., 1950, 'Personality and language in society'. In: Papers in Linguistics 1934-1951. Oxford U.P.
Firth, J.R., 1951, 'General linguistics and descriptive grammar'. In: Papers in Linguistics 1934-1951. Oxford U.P.
Firth, J.R., 1957, 'A synopsis of linguistic theory 1930-55'. In: Palmer, ed.
Fishman, J.A., 1971, 'The relationship between micro- and macro-sociolinguistics in the study of who speaks what language to whom and when'. In: Pride and Holmes, eds.
Gimson, A.C., 1973, An introduction to the pronunciation of English. Edward Arnold.
Gregory, M., 1967, 'Aspects of varieties differentiation '. Journal of Linguistics 3.
Gregory, M., Carroll, S., 1978, Language and situation. Routledge & Kegan Paul.
Halliday, M.A.K., McIntosh, A., Strevens, P., 1964, The linguistic sciences and language teaching. Longman.
Halliday, M.A.K., 1970, 'Language structure and language function'. In: Lyons, ed.
Halliday, M.A.K., 1973, Explorations in the functions of language. Arnold.
Halliday, M.A.K., 1978, Language as social semiotic. Arnold.
Halliday, M.A.K., 1980, 'How is a text like a clause?'. Nobel Symposium on Text Processing. Stockholm.
Holland, V.M., Redish, J.C., 1982, 'Strategies for understanding forms and other public documents'. In: Tannen, ed.
Huddleston R., Hudson, R.A., Winter, E.O., Henrici, A., 1968, Sentence and clause in scientific English. Osti Report 5030, University College London.
Hymes, D., 1971, 'On communicative competence'. In: Pride and Holmes, eds.
Kittredge, R., Lehrberger, J., eds., 1982, Sublanguage. Studies of language in restricted semantic domains. Walter de Gruyter.
Lee, W., 1969, 'Some points about aims and means in the foreign-language course'. English Language Teaching . Vol. 23/2
Leech, G., 1966, English in advertising: A linguistic study of advertising in Great Britain. Longman.
Leech, G., Svartvik, J., 1975, A communicative grammar of English. Longman.
Leech, G., Short, M., 1981, Style in fiction. A linguistic introduction to English fictional prose. Longman.
Lyons, J., ed., 1970, New horizons in linguistics. Pelican.
Malinowski, B., 1923, 'The problem of meaning in primitive languages'. Supplement to Ogden and Richards, The meaning of meaning. Kegan Paul.

Martin, J.R., 1981a, Register and meta-function. Mimeo. University of Sydney. Department of Linguistics.
Martin, J.R., 1981b, Lexical cohesion. Mimeo. University of Sydney. Department of Linguistics.
Martin J.R., 1981c, 'How many speech acts?' University of East Anglia Papers in Linguistics. 14-15.
Martin, J.R., 1981d, The development of register. Mimeo. University of Sydney. Department of Linguistics.
Martin, J.R., 1981e, Conjunction: the logic of English text. Mimeo. University of Sydney. Department of Linguistics.
Mitchell, T.F., 1957, 'The language of buying and selling in Cyrenaica: a situational statement'. Hesperis.
Munby, J., 1981, Communicative syllabus design. Cambridge U.P.
Nash, W., 1980, Designs in prose. Longman.
Palmer, F.R., ed., 1968, Selected papers of J.R. Firth 1952-1959. Longman.
Perren, G.E., Trim, J.L.M., eds., 1971, Applications of linguistics. Cambridge U.P.
Pett, J., 1982, 'Reading: an ESL skill at the university'. Forum XX/3
Pride, J.B., Holmes, J. eds., 1971, Sociolinguistics. Penguin.
Reid, T., 1956, 'Linguistics, structuralism and philology'. Archivum Linguisticum 8.
Simon-Vandenbergen, A.M., 1982-83: 'News reports and comments'. Studia Germanica Gandensia, XXII.
Simon-Vandenbergen, A.M., 1984, The semantics of political press reportage: a quantitative approach'. Communication & Cognition, Vol. 17, n° 4.
Sinclair, J., Coulthard, R., 1975, Towards an analysis of discourse. Oxford U.P.
Strevens P., 1977, New orientations in the teaching of English. Oxford U.P.
Tannen, D., ed., 1982, Analyzing discourse: text and talk. Georgetown U.P.
Taylor, C.V., 1983, 'Structure and theme in printed school text'. Text 3/2.
Ure, J., 1968, 'Practical registers'. English Language Teaching, 23/2-3.
Ure, J., 1971, 'Lexical density and register differentiation'. In: Perren and Trim, eds.
Ure, J., 1982, 'Introduction: approaches to the study of register range'. International Journal of the Sociology of Language 35.
Vandamme, F., 1977, "Register-semantics'. Communication & Cognition vol. 10.
Vandamme, F., 1981, 'The combination of registers. A non-atomistic approach'. Werkgroep 'Pragmatiek en onderwijs', Rijksuniversiteit Gent. Vol. 36.

Vandamme, F., 1982, 'De dicto, de re'. Werkgroep 'Registers, kennisrepresentatie, kennisverwerving en -verwerking', Rijksuniversiteit Gent, Vol. 50.

Vervenne, D., 1983, 'Concerning analogical reasoning: a register-pragmatic approach to classroom and laboratory contexts'. Paper presented at the 3rd Language and Language Acquisition Conference. Ghent 21-25 March 1983.

White, R., 1974, 'Communicative competence, registers and second language teaching'. Iral XII.

Widdowson, H., 1973, 'Directions in the teaching of discourse'. In: Corder and Roulet, eds.

CHILDREN'S IDEAS ABOUT THE FORM AND FUNCTION OF QUESTIONS

Anne Sinclair

University of Geneva

1211 Genève 4 Switzerland

INTRODUCTION

Questions

Everybody knows what a question is. Linguists, psycholinguists, logicians, as well as laymen hold the same definition of 'questions', though the professional definition may of course be more rigorous, more complete, more coherent. We conceive of questions as being written or verbal sequences that have certain grammatical properties; they belong to the class of utterances called 'interrogatives'. For example, Belnap and Steel (1976, p. 201) in the index to their book The Logic of Questions and Answers say "to avoid needless repetition, we assume an automatic crossreference from 'interrogative' to 'question'" [1].

Generally speaking, for both laymen and scholars, non-verbal questions (raising the eyebrows for example) and introspective questions are excluded from the class of questions. Further, when considering questions, it is obviously important not to lose sight of answers.

The fact that a strong, often implicit concensus exists as to the meaning of the word 'question' is reflected in the psycholinguistic literature on the topic. Authors usually do not give a definition of what a question is, since they take the implicit definition for granted. For example, Robinson and Arnold (1977) do not describe or discuss questions themselves: no matter, readers know what the topic is. Commonly, authors in the psycholinguistic terrain concern themselves with a sub-set of the class of questions and the subset is easy to define; for example, Cairns and Hsu

(1978, p. 477): "... children ... were asked six types of wh-questions following video-taped sequences". Some authors, such as Wode (1971) do give exhaustive definitions of the class of utterances that they are studying [2].

Although definitions and the setting up of taxonomies are not always the most important preliminary steps, the lack of them may often rapidly lead to problems. In psycholinguistics, when dealing with early child language, it is evident that implicit definitions (or elaborate linguistic definitions) must be reviewed or challenged, and new definitions set up. Are we to call "Mama" said with a rising intonation a question? Or is it a proto-question? Or is it a request? Contextual information and knowledge of the child and his speech habits will have to be taken into account to discover his underlying intentions. The analysis and classification of the utterance "Mama" will in some way depend on these inferred intentions. Although this problem has been carefully handled (see Dore, 1975), the only solution is to go beyond implicit definitions and definitions of form and to attend to underlying intentions and functions.

The second difficulty linked to a formal definition of questions arises in the interpretation, not of utterances themselves, but of psycholinguistic results. Working within an implicit or explicit formal grammatical definition of questions, developmental psycholinguists have outlinded how (in various languages) children gradually come to produce and understand different question forms (see Brown, 1968; Klima and Bellugi, 1966; Ervin-Tripp, 1970; Wode, 1971, Stewart and H. Sinclair, 1975; Cairns and Hsu, 1978; Tyack and Ingram, 1977). Rough orders of acquisition emerge. For example, 'what' and 'where' are acquired before 'who', 'why' and 'how', with 'when' appearing last. To explain this type of development, researchers find they have to refer to meaning. For example: "It has been assumed that what and where questions are learned earliest because they represent non-linguistic notions (identification and location of objects) which are developed quite early. Conversely, when questions are acquired late because they are based on concepts of time which the child acquired at a later stage". (Lightbown, 1980, p. 153) Goffman (1976, p. 257) puts the point neatly: " ...the meaning of an utterance, whether question or answer, can ultimately depend in part on the semantic value of the words it contains and thus (in the opinion of some linguists) escape complete formalization". The problem of the interdependency of form and meaning is traditional in linguistics and psycholinguistics.

In developmental psycholinguistics, a certain trend deals with the content (meaning) of questions, rather than their form or their function. The semantic analysis of questions (called the descriptive analysis of question types by Kearsley, 1976) has attracted little attention and seems to have gone slightly out of fashion, perhaps because it involves analyzing a wide-range of spontaneous production data. Piaget (1923) was the innovator and subsequent studies (Davis, 1932; Meyer and Shane, 1973) have confirmed the pertinence of his analysis and added larger span developmental data. Piaget analyzed all the questions produced by one child (Del) during a ten month period. His analysis of Del's questions concentrates on content (questions about causality, about reality and history, questions about actions, intentions, rules, etc.). Piaget relates Del's questions to the child's cognitive development.

Apostel (1981, p. 23) after pointing out that questions fulfill different functions in dialogue (compare "Could you pass the salt?" with "What time is it?") states that "une théorie purement syntaxique de l'interrogation n'est donc pas possible" (a purely syntactic theory of interrogation is illegitimate). Interrogative sequences produced in dialogue have a great variety of function, illocutionary forces and perlocutionary effects (see Austin, 1962; Searle, 1969). "Could you close the door?" is a request and perhaps does not even require a verbal response. "Who cares?" is rhetorical. "How many times have I told you not to squeeze the toothpaste in the middle?" is a reproach of some kind. "Why not go to Mexico?" is a suggestion, etc. Functional criteria and classifications have preoccupied many theorists and researchers.

Fauconnier (1981) shows that it is impossible to treat questions like those above (requests, commands, etc.) in the true sense of 'interrogation'. What is true 'interrogation' from a functional perspective? Both layman and scholars feel that the prototype question (see Flammer, 1980) is an information request. People ask questions to obtain new information, to resolve doubt, to overcome uncertainty. "How do airplanes fly?"; "What time does the bus leave?"; "Are you sure Joe is coming?"; "Do you want coffee or tea?" ... these are 'real' questions.

The concept of information question hinges, of course, on considerations concerning the knowledge (or degree of uncertainty) of the speaker and the hearer. Without being exhaustive or precise, speech act theorists generally make some of the following assumptions: S does not know (or is uncertain about) what he is asking; S supposes that H will not tell him unless he asks; S presumes that H does know, etc. Nevertheless, 'information questions' (the class of questions generally treated by speech-act theorists) remains ill defined. Some question types share some (but not all) of the properties of information questions, and these are included or excluded, discussed or not. Human beings like to ask questions to which everyone knows there is no answer ('How many angels can dance on the head of a pin?"); they ask questions to which they already know the answer ("How much is two and two?"), and many of the questions asked during dialogue refer to the dialogue itself, and not to other cognitive or extralinguistic matters ("What did you say?").

Lyons (1977) reviews the work concerning what we must loosely and globally call 'information questions'. Speech act theorists generally classify them as one of the subclasses of a category called mands, along with commands and requests. In their turn, mands are a sub-class of directives. Briefly, directive utterances suggest or propose some future course of action or pattern of behavior (demands, warnings, suggestions, orders, etc.) for the addressee. Their felicity condition is that S must believe that H can comply. Mands (see Ross, 1968) are those directives where S wants H to comply (which need not necessarily be the case, for example, when one is making a suggestion). Questions are thus seen as instructions to make a statement, in analogy to commands and requests which are instructions to carry out actions or to behave in a certain way (see Hare, 1949; Lewis, 1969). Questions are seen as implicitly meaning "bring it about that I know that such and such is so" (Acqvist, 1965; Hintikka, 1974). Seen in this way, information questions neatly fit the category of mands. Lyons himself has several criticisms of this analysis and proposes an alternative one, i.e. to make a distinction between asking a question and posing a question. When we pose a question we "merely give expression to, or externalize our doubt... and we can pose questions which we do not merely expect to remain unanswered, but which we know or believe to be unanswerable" (p. 755). All the examples he gives are indirect questions (declaratives which contain an embedded partial interrogative phrase: "I wonder whether the door is open"). The difficulty of combining functional and formal criteria is once more apparent: Speech act theorists define questions grammatically (though this part of the definition is often implicit!), analyze or concentrate on a sub-class of these according to functional criteria, and find themselves having to extend or change the formal definition (for example, 'questions' may include 'indirect questions', as described above).

Discourse analysts present models that are based on more broadly functional criteria, and not on implicit criteria of form. J.M.cH. Sinclair and Coulthard (1975) define the discourse value of their main initiating acts as: elicitation (H's subsequent act: reply), directive (react) and informative (acknowledge). The function of elicitation is to request a linguistic response. It is often realized by questions, but may be realized by other means. Questions can fulfil other function in discourse (commands, etc.) and may even serve to make a response, as in: "Have you seen my coat?" (elicitation) "Isn't that it over there?" (response).

Work in child language undertaken from a functional or speech act perspective reflects these slightly devergent view-points.

Garvey (1975), for example, set out to study how young children make requests for actions. A small portion of the requests in her data are indirect requests in the interrogative form ("Can you hold the baby?"). Others, rather than studying how a certain functional communicative aim is expressed verbally, have investigated the functions fulfilled by interrogative sequences in the child's discourse. Holzman (1972) shows that children in the three morpheme period already use the interrogative form for making suggestions and reports (as their mothers do). Using a similar approach, but working with older children, Van Hekken and Roelofsen (1982) classify the interrogative sequences collected into various categories: knowledge questions of various types, questions asking for permission, soliciting agreement or attention, etc.

To sum up, interrogative sequences are easy to define grammatically. Once we are dealing with the function of utterances, we note that criteria of form are not adequate, or rather that the relationship between form and function is complex and indirect, and that little is known about it. Lastly, from a broadly functional perspective, very little is known about why and how human beings elicit verbal responses from each other.

Children's Ideas

Having given a brief review of the ideas of psycholinguists (and others) on questions, let us consider the pertinence of studying children's ideas on the topic.

Over the past twenty years, interest in the child's meta-linguistic abilities (or linguistic awareneness) has been growing (see A. Sinclair et al., 1978; Hakes, 1980). Many different types of language behavior and use have been studied, as well as syntax and semantics. Clark (1978) gives a good review of the field. Meta-linguistic studies have been performed with a variety of explicit and implicit theoretical interests. They have sprung from concerns with meta-cognition, interest in the psychological reality of grammatical transformations, and, naturally, speculations about the links between development or acquisition and awareness. Last but not least, awareness of different aspects of language is a topic of study in its own right (see H. Sinclair, 1978). This is the point of view adopted in this paper. I do not expect that tracing the development of children's ideas about questions will directly inform us about the links between conceptions and behavior, or indeed, how comprehension and production develop.

Why were questions chosen? Questions are psychologically and conversationally important. 'Question' is a frequent and well-known word, although it is likely that the meaning young children attach to it is not the same as that of adults. Questions are also produced early by children, and a large proportion of the utterances they hear and respond to are questions. In short, questions seemed a fruitful terrain for trying to grasp the development of ideas about form and function, and how they are articulated.

METHOD

We asked 65 French-speaking children aged 5 to 12 (5 children of 5, and 10 children of each of the following age groups: 6, 7, 8, 9, 10, 11-12) to perform the following tasks [3].

A. In a meta-linguistic discussion with the child, we asked him to define what a question was. We asked: "What is a question? Can you give examples? How do you know that that is a question? How can you tell if something said is a question or not? Is 'Bonjour Madame' a question? If so, how do you know? If not, why not? (similar judgments were obtained for other utterances)". We also asked: "Can you say something which is not a question? When does one ask questions? Why? Who asks questions?". This part of the study was conducted as a clinical interview. The experimenter took up the child's responses, gave counter-examples, etc. For example, if the child said "A question is when you have to give an answer", the experimenter also asked him about answers.

B. We asked the child to play the role of a child facing a teacher (half the subjects) or a teacher facing a child (the other half). Children were given a puppet and the experimenter took another puppet. The child was asked to say ("dire" or "poser", never "demander") many questions, or only questions, if he could, to the child or teacher. The experimenter answered the child's questions in a natural manner, but avoided asking questions in return, thus putting the burden of conversation-management on the child. We analyzed the first ten utterances (defined as turns) produced by the child (or less, if less was produced).

C. The children were asked to produce questions with paticular question words. This was done by presenting them with three example utterances for each question word (for example, for why: Why are you wearing a sweater? Why do we have to eat? Why don't you come play with me?), produced by another puppet, and asking them to "say things like the puppet". If instructions were not followed as we wished, we specifically asked the children to "say something with why". Question words used were pourquoi, quand, où and qui (why, when, where, who(m)).

We thus hoped to obtain information concerning the child's theoretical ideas about the form and function of questions, as well as his implicit criteria of form and function (to produce a lot of questions). Tasks B an C will permit us to analyze the form and content of the questions produced by the children (these analyses are not reported in this paper). Finally, the C task tests the epi- and meta-linguistic capacities.

RESULTS

Our tasks were too difficult for the 5 year-olds. In the A task, they defined questions by saying "A question, that's saying something". All utterances (except meaningless ones like bababababa, which we also presented for judgment to our younger subjects) were said to be questions. The examples they gave were all commands ("Go to bed!").

Difficulties with the role playing situation (B task) resulted in two different types of dialogues: either the children ignored the puppet and spoke as themselves to the experimenter, or they played fantasy games with the puppets.

5 year-olds took more readily to the C task, sometimes producing questions with the different question words. (The results of the C task are given in Table 1).

However, they often began to speak (as in ordinary conversation) to the experimenter, or persevered with one question word (why) when asked to produce questions with another question word.

At the age of 6, the conception of questions implicit in the 5 year-olds responses became more clearly articulated, but hardly changed. For task A, 7 of the 10 children give definitions like the following: "It's when you speak"; "One says things because one can say them, because one wants to say them"; "Questions is when you say things to people, when people tell each other things" [4]. These children accepted all utterances as 'questions', provided they had meaning. The examples they gave, however, were always commands (imperative form) or requests [5] for future actions. For example: "Go and get the newspaper", or in reported discourse, "my father says you must ('il faut') eat, you must drink a glass of milk". The commands quoted all seemed to be examples of parental speech and pertained to future actions to be performed by the child, generally related to daily routine (dressing, eating, sleeping ...). The 3 other children aged 6 gave answers of a higher level. One of them said "One asks questions to know life better" and gave examples like the following: "If one asks what is your name", "She asks what do you want". Another child associates questions with polite requests: "If one finds something one wants, one asks one's father, if one can buy it". Children aged 6 could not distinguish between questions and non-questions and continually hesitated and changed their minds, in accord with criteria that remained mysterious to us. Typical answers are of the type:

" 'Bonjour Madame' is a question that I say".

Most of the 6 year-olds could not perform our B task, and either behaved like the 5 year-olds or explicitly told us that they did not know what to say. One child produced two utterances: "Why does one have to write? (R)[6] Why does one take a piece of paper? Well I'm stopping now". Another child produced only commands in the imperative form: "Go to the janitor ... Go and sit on the bench ... And now ...", etc. Only one child produced several consecutive utterances in the interrogative form, addressing the puppet and not a hypothetical teacher ("Why do you have black hair? (R) Why do you have blue eyes?").

For the C task, Table 1 shows that several children had no difficulty in producing questions with why, the other question words being more difficult. When they did not succeed in producing a series, they either continued to ask questions with why, or asked questions, but of a different type ("Do you like swimming?"; "Is the coffee good?", for example), or said nothing at all.

At 7 years of age, three children gave the same type of responses as those aged 6. The other children's responses in task A showed that their conception of questions was more elaborated. Several new elements appeared. Firstly, questions are seen as being accompanied by some kind of response (not necessarily a verbal one). For example: "When they say things and you must answer"; "Saying something and afterwards somebody answers"; "They tell you and then you must do it". Secondly, it is clear that for these 7 year-old children, 'question' is associated with requests for objects (which may also be formulated as commands, see footnote 5), or requests for permission to carry out certain actions. For example: "You ask questions, because often you don't have the things you need".

The examples of questions they gave bear out this conception, and are requests for objects formulated as questions, assertions, or imperatives: "Hey, Mommy, can I have a piece of cheese?"; "Give me the dog"; "Can I please have a sweet?"; "I'd like an ice-cream, a piece of cake"; or, a request for permission; "Can I go outside?"

In the B task, these children of 7 did role-play adequately and asked many questions. They tended to construct dialogues that are very life-like in which almost all the utterances are directives. One typical example: "What do you want to do drawing, or pottery? (R) What do you want to do? I have to know that first. You're going to draw a house. Go and get a piece of paper in the cupboard. You can do what you like, but there must be a house, while I'm working. Can I see? (R) What do you want to do? (R) Take your bag. We're going down. What do you want to play? Basketball, football? (R) How much is five and five? (R) How much is eight and eight? ..."

In the C task, 7 year-olds found it easier to produce questions with the different question words. If they did not do so, they asked questions of different types.

At the age of 8, in the A task, the idea that questions serve to obtain information appeared in various forms. "One asks questions when one does not know things"; "I say, for example, where do you live? Because I don't know it"; "Because you want to know things". Questions were also intimately linked to answers, but clearly distinguished from them. At 8, children often point out that answers are not always obtained. Strangely enough, the examples the children gave did not bear out these early intuitions of the information-obtaining function of questions; just like with the 7 year-olds, almost all the examples given were requests for objects, with requests for permission to carry out certain actions becoming somewhat more frequent. ("Can I have a piece of cake?"; "Can I go to Mary's?" ...) These utterances were not all in the interrogative form, and only one child gave examples of questions that clearly had an information-obtaining function (he gave a whole series). Our examples of non-questions were rejected as questions, but without clear justification. For example: "'Bonjour Madame', that's not a question, because I don't say so many things"; " 'Bonjour', that's not a question, because you don't say what is a 'bonjour' ".

In the B task, at the age of 8, most of the children produced only interrogatives. They often did so by picking one model and varying certain words. For example: "What is a dictionary?" ... What is orangeade? What is ...", etc. Others realistically produced questions in the role of teacher or child. "Hello, what do you want to do? (R) Do you want to play, or fill in the work-sheets, or work? (R) No, OK, so what is ten and four? (R) Bravo, good and five and twenty? ..."

Table 1 suggest that it is rare for an 8 year-old to be incapable of producing a series of questions with the various question words (C task)

The children of 9 all defined the function of questions (A task) as being that of obtaining information. "A question is something you ask someone when you don't know something"; "It's asking somebody something to know it or to get information". The interlocutor was specifically mentioned, and the speaker's knowledge (or rather, lack of it), played an important role. The examples given confirm this conception : "Where did Eric get to?"; "When is your birthday?"; "Are these curtains already sold?"; "Who invented electricity?", etc.

When these children were asked how one might recognize or define questions, they all gave criteria of form. Each child gave

Table 1. Number of children who pruduce a series of questions with the different question words.

AGE	why	where	when	who(m)
5 (n=5)	4		1	
6 (n=10)	6	2	2	1
7	7	5	6	2
8	9	10	10	10
9	10	10	10	10
10	10	10	10	10
11-12	10	10	10	10

at least one of the following criteria, and many gave several of them:

a) an answer is called for or necessary
b) a yes/no answer is possible (or necessary)
c) subject and verb are inverted
d) particular words (questions words) and phrases ('est-ce que...?') are present
e) the utterance has a particular intonation (generally, rising at the end).

They naturally find subject-verb inversion difficult to express. "Some words are turned around"; "Hum, the subject. The verb ... the verb in front of the subject"; "The pronoun and the verb are different". However, when rejecting non-questions they generally justified their judgments with "one isn't asking", or "no answer is needed" ... speech-act type and dialogic components played a larger role than criteria of form.

From the age of 9 up, our B task is easy for the children. All of them produced only interrogatives. Many of the older subjects did not even role-play a dialogue (i.e. did not wait for responses), but simply fired off a list of questions of various types. Table 1 shows that task C was sucessfully performed by all children aged 9 and older.

At the age of 10, we obtained responses similar to those

given by the children of 9 in the A task. Some of the 10 year-olds add extra points: questions are not only to ask something one does not know but that one would like to know, and one may ask questions not to find out something, but to find out if the other person knows. The formal characteristics of interrogative sequences were the same as those mentioned by the 9 year-olds, but each child mentioned more of them. Non-questions were rejected according to criteria of form, and not according to dialogic criteria.

The oldest group of subjects (11-12) produced the same type of responses. They were more coherent, more consistent and elaborate in their descriptions, judgments and explanations: their conception of questions had not changed, but their capacities for arguing and explaining seemed greater.

DISCUSSION

Our results show that the child's conception of the definition of and the function of questions go through several stages. It is clear that children base their ideas not only on the meaning of the utterance, or, at a later age, on their form, but that their illocutionary force or their function in dialogue plays a large role. It is surprising that the results obtained fit so easily into a speech-act or discourse perspective.

Although many young children (age 5, 6 and 7) define questions as being any utterance at all that has meaning, the examples given and their behavior in the B task shows that the following development occurs. At first, 'question' is associated with commands (or requests) for future actions (usually, no objects and no other person than the addressee are involved). Next, 'questions' are associated with utterances that are requests (or commands) for objects. At the same time, requests for permission to perform certain actions are also given as examples, and these are the first utterances quoted that fulfill a purely verbal function (the addressee, usually a parent, gives his verbal permission for certain actions to be carried out). Lastly, at the age of 9 and older, the function of questions is seen as obtaining information, and the function of the utterances quoted is to obtain a verbal response only. After the age of 9, information questions are not only seen as the 'prototypical question', but the link between obtaining information and different types of interrogatives is univocal. This point of view is in accordance with a certain tradition in linguistics and psycholinguistics; naturally, the prevailing opinion is reflected in what children are taught (and in this case, learn) in school. To our mind, it seems likely that information questions have certain particularities that neither children, speech act theorists nor linguists have as yet uncovered.

The development is difficult to interpret. It was unexpected, and no similar results exist in the literature, to my knowledge. One might say simply that young children do not understand the word 'question' and answer somewhat randomly, in order to satisfy the experimenter. While this may be true in a certain sense, it cannot serve as an explanation, for the following reasons. Firstly, children answered naturally -- no child refused to answer, or told us that he did not know what a question was. Secondly, children of different ages respond to out tasks in a consistent fashion. For example, no child produced a simple assertion or representative (such as "The sky is blue" or "Mary gave me this bracelet") as an example of a question. Thirdly, the conceptions the young children have seem to be personally constructed -- one cannot imagine parents or teachers telling children that commands for actions are 'questions' or in any way related to them.

Any interpretation of these results should take into account the fact that children's conceptions of utterance-type are dependent on interactional value and not on form. We might hypothesize that for young children 'question' is associated with a striking, visible or important perlocutionary effect, involving actions and objects: a kind of utterances-as-actions perspective. A description of the results in these terms would continue as follows: it is only around the age of 9 that children can conceive of utterances on a purely verbal or discourse level, one utterance "calling forth" another, without any other links with the activities of the participants.

Other interpretations are also possible. One might say that in commands and requests the meaning-intention of the speaker is particularly clear to the addressee, and is therefore easy to reflect on. In children's minds, these kinds of utterances would thus have a particular salience. Such an interpretation might continue as follows: it is only around the age of 9 that children can conceive of a speaker knowing that he does not know something and that someone else might know what he does not know, and that he can thus obtain the desired information from somebody else.

The results do not readily permit us to decide which of these interpretations (and the many others that are possible) best fit the facts.

As far as the form of questions is concerned, children below the age of 8 - 9 do not explicitly mention it. The first formal criteria that appear are dialogic and are linked to the fact that question and answer form an adjacency pair. At the ages of 7, 8 and 9 questions are often principally defined by the fact that they demand a response. Children of 9 and older will mention formal, grammatical criteria. No erroneous criteria are given, though definitions are sometimes partial and children often have great difficulty in expressing themselves. The results regarding form are relatively easy to interpret. Research in the field of metalinguistic awareness has often shown that awareness of form appears well after a certain form has been acquired (production and/or comprehension). Research has also shown that young children find it difficult to consider utterances independently of their meaning or the context in which they are produced. Thinking about language as an object that has general characteristics of a formal nature is a particular cognitive activity that requires a great deal of decentration.

As for the links between function and form, we may note that three types of behaviors seem to be linked, in that they appear at the same ages: a) considering the function of questions to be an information obtaining one; b) being able to give at least some formal characteristics of interrogative sequences; c) being able, in the B task, to produce only interrogatives, which implies having clear implicit ideas about form, and using these ideas to control production. Exactly how ideas about form and function interact merits further study.

ACKNOWLEDGMENTS

The research described in this paper was carried out thanks to the Fonds National de la Recherche Scientifique, grant no. 1-965.-0.79. I am grateful to M. Croisier, R. Koyoma, L. Marini and C. Ruata for help with the study.

FOOTNOTES

1. Throughout this paper, question is used in this sense, unless specified otherwise.
2. Descriptions of questions vary according to the language that is being considered (see Greenberg, 1963 for an overall survey of question forms). Segmentally marked questions contain an interrogative particle or pronoun. In English these are termed wh-questions (who, what, when, etc.).
   Word order questions: in English and French, this order always takes the form of subject-verb inversions of various types.
   Intonation questions: only a particular intonation shows that a question is being asked. Semantically, questions are divided into open questions (wh-questions) and closed questions, questions that may be answered with yes or no.
3. The study reported was carried out in French. Briefly, questions in French may be formed by:
   a) subject-verb inversion (simple inversion). Or, with the use of an interrogative pronoun, adjective or adverb: Quand Pierre vient-il? Pourquoi Marie est-elle malade?

b) by the use of est-ce que/qui and subject-verb order.
c) by the simple juxtaposition of an interrogative word and an affirmative phrase. Où vous habitez? Vous habitez où?
d) by using intonation only: Pierre est malade?
Open questions are called interrogation partielle (or: nucléaire) and closed questions are called interrogation globale or totale.

4 . These comments are all translations; some of them may seem untypical of child speech.

5 . Setting up rigorous distinctions between commands and requests is not easy (one must consider form, directness, politeness, status of the interlocutors, etc.). In the framework of this study, we did not attempt to do so, because we were dealing with examples constructed by children, out of context. The distinction between commands for actions and requests for objects, while reposing on semantic criteria, was nevertheless seen to result in rough distinctions dealing with form: imperatives for commands and interrogatives and assertions, as well as politeness markers, for the requests.

6 . (R) indicates that the experimenter made an appropriate response.

REFERENCES

Apostel, L., 1981, De l'interrogation en tant qu'action. Langue Française, 52, 23-43.

Aqvist, L., 1965, A new approach to the logic of questions. Philosophical studies, University of Uppsala.

Austin, J.L., 1962, How to do things with words. Clarendon Press, Oxford.

Belnap, N.D. and Steel, T.B., 1976, The logic of questions and answers. Yale University Press, New Haven, London.

Brown, R.W., 1968, The development of wh-questions in child speech. Journal of Verbal Learning and Verbal Behavior, 7, 279-290.

Cairns, H., and Hsu, J.R., 1978, Who, why, when and how: a developmental study. Journal of Child Language, 5, 477-488.

Clark, E.V., 1978, Awareness of language: some evidence from what children say and do. In A. Sinclair, R.J. Jarvella and W.J.M. Levelt, eds., The child's conception of language. Springer, New York.

Davis, E.A., 1973, The form and function of children's questions. Journal of Genetic Psychology, 123, 285-296.

Dore, J., 1975, Holophrases, speech acts and language universals. Journal of Child Language, 2, 21-40.

Ervin-Tripp, S., 1970, Discourse agreement: how children answer questions. In J.R. Hayes, ed., Cognition and the Development of Language, Wiley, New York.

Fauconnier , G., 1981, Questions et actes indirects. Langue Française 52, 44-55.

Flammer, A., 1980, Towards a theory of question asking. Research Bulletin, no. 22, University of Fribourg.

Garvey, C., 1975, Requests and responses in children's speech. Journal of Child Language, 2, 41-63.
Goffman, E., 1976, Replies and responses. Language in Society 5, 257-313.
Greenberg, J.H., 1963, Some universals of grammar with particular reference to the order of meaningful elements. In J.H. Greenberg, ed., Universals of Language, MIT press, Cambridge, Mass..
Hakes, D.T., 1980, The development of meta-linguistic abilities in children, Springler, New York.
Hare, R.M., 1949, Imperative sentences. Mind, 58, 21-39.
Hintikka, J., 1974, Questions about questions. In M.K. Munitz and P.K. Unger, eds., Semantics and Philosophy. New York University Press, New York.
Holzman, M., 1982, The use of interrogative forms in the verbal interaction of three mothers and their children. Journal of Psycholinguistic Research, 1, 311-116.
Kearsley, G.P., 1976, Questions and questions asking in verbal discourse: a cross-disciplinary review. Journal of Psycholinguistic Research, 5, 355-375.
Klima, E.S., and Bellugi, U., 1966, Syntactic regularities in the speech of children. In J. Lyons and R.J. Wales, eds., Psycholinguistic Papers, Edinburgh University Press, Edinburgh.
Lewis, D., 1969, Convention. Harvard University Press, Cambridge, Mass.
Lightbown, P.M., 1980, The acquisition and use of questions by French L2 Learners. In S.W. Felix, ed., Second Language Development, Gunter Narr Verlag, Tübingen.
Lyons, J., 1977, Semantics (vol. 2), Cambridge University Press, London.
Meyer, W.J., and Shane, J., 1973, The form and function of children's questions. Journal of Genetic Psychology, 123, 185-296.
Piaget, J., 1923, Le Langage et la Pensée chez l'enfant, Delachaux et Niestlé, Neuchâtel.
Robinson, W.P., and Arnold, J., 1977, The question-answer exchange between mothers and young children. European Journal of Social Psychology, 7, 151-164
Ross, A., 1968, Directives and Norms. Routledge and Kegan Paul, London.
Searle, J.R., 1969, Speech acts: An essay in the philosophy of language. Cambridge University Press, New York.
Sinclair, A., Jarvella, J.R. and Levelt, W.J.M., 1978, The child's conception of language. Springer, New York.
Sinclair, H., 1978, Conceptualization and awareness in Piaget's theory and its relevance to the child's conception of language. In A. Sinclair, R.J. Jarvella and W.J.M. Levelt, eds., The child's conception of language, Springer, New York.
Sinclair, J., McH. and Coulthard, R.M., 1975, Towards an analysis of discourse: the English used by teachers and pupils. Oxford University Press, Oxford.

Stewart, J., and Sinclair, H., 1975, Comprehension of questions by children aged between 5 and 9. International Journal of Psycholinguistics, 3, 17-26.

Tyack, D., and Ingram, D., 1977, Children's production and comprehension of questions. Journal of Child Language, 4, 211-224.

Van Hekken, S.M.J., and Roelofsen, W., 1982, More questions than answers. A study of question-answer sequences in a natural setting. Journal of Child Language, 9, 445-460.

Wode, H., 1971, Some stages in the acquisition of questions by monoligual children. Word, 27, 261-310.

INDIVIDUAL LANGUAGE AWARENESS TESTING AND EARLY READING

Marc Spoelders and Lut van Damme

Laboratorium voor Pedagogiek

Rijksuniversiteit Gent

1. INTRODUCTION

Developmental psycholinguistic research makes it abundantly clear that in normal circumstances, given an educationally adequate support, young children make pronounced progress in the use of language. When they are to learn to read, their language production and comprehension abilities seem to have reached a point from which reading, too, would seem an attainable undertaking. Too many children, however, lose this battle. Traditionally, educationalists have paid much attention to perceptual and motor factors for explaining reading failure. However, a relatively important portion of reading failure does not seem to be based on malfunctioning in these areas. The hypothesis is advanced that reading requires also metalinguistic awareness, the reflection on language and linguistic behaviour as objects of thought, as things that could be analyzed. Although the precise nature of the relationship between metalinguistic awareness and the acquisition of reading skill has not yet been revealed recent theorizing in this domain attributes explaining value to the former (for a review, see bibliography). Based on these theoretical considerations we formulate the following hypothesis:

> Children who have a higher level of metalinguistic awareness when they begin learning to read, will achieve better in reading than children who manifest a lower level of metalinguistic awareness.

In order to measure metalinguistic awareness we need a valid and reliable instrument. Since no such device (in Dutch) is at the disposal of psycho-educational researchers we had to construct one of our own. We decided that a first version of it should contain a large pool of items so that after the try-out a shorter, eventually less time-consuming one could be constructed. The longer version is called the parent-instrument (PI). Its different parts and items have been selected and adapted from the range of materials reported in psycholinguistic literature on metalinguistic awareness. We hope the final version will develop into a useful research instrument for psycho- educational purposes, as well as an informative psychodiagnostic device for teachers and educationalists alike.

In what follows we describe the PI and the (try-out) sample. We also discuss the results of a first investigation into the relationship between metalinguistic awareness and reading achievement. Finally we report the steps that led to the construction of a shortered instrument, called GTL (Geïndividualiseerd Taalbewustzijnsonderzoek bij het Leesbegin / Test on Individualized Language Awareness at the beginning of Reading Instruction).

2. STRUCTURE OF THE PI

The PI contains three main sections (see Table 1)

I The first part of this instrument concerns the recognition of literacy behaviour and the understanding of literacy functions. Recognizing activities such as reading and writing exemplifies a sort of general awareness (I1). In our opinion, it may be very difficult for a child to learn to read if it has no working definitions of reading and writing. Some investigations conclude that, in fact, many first grade pupils do not have any specific expectancies on what reading is going to be like, or what it consists of.They also indicate that young beginners have difficulty in understanding the purpose of written language (I2), in acquiring a kind of pragmatic awareness. The PI tests the child's knowledge of these aspects by means of drawings that either do or do not illustrate reading or writing activities, or the purposes of literacy. For example:

- The child is shown a picture of a girl writing her name down on the plaster of the broken arm of her friend, and is asked 'What is this girl doing?'

- The child is shown a picture of a man reading the method of preparation on tinned soup and is asked 'Show me everything that people can read. Point to it with your finger.'

- The child is shown a picture of a man sitting in his car reading the sign posts and is told 'This man has lost his way. Can he find the right road this way, as on the drawing? How do you know?'

Table 1: Structure of PI and GTL.

	PI	material	items	GTL	material	items
I	1 Recognizing literacy behaviour	drawings	32	1 Recognizing literacy behaviour	drawings	10
	2 Understanding literacy behaviour	drawings	28	2 Understanding literacy behaviour	drawings	20
II	Concepts of features of printed language	test		Concept of features of printed language	test	
	Knowledge of technical terminology	booklet	36	Knowledge of technical terminology	booklet	20
III	Concepts of features of spoken language			Concepts of features of spoken language		
	Phoneme awareness			Phoneme awarenass		
	1 Phoneme differentiation	2 frogs	12	1 Phoneme differentiation	2 frogs	6
	2 Rhymes	puppet J	12	2 Rhymes	puppet J	6
	Word awareness			Word awareness		
	3 Concept word		12	3 Word boundaries	cubes	6
	4 Long / short words		12	4 Long / short words		6
	Form awareness			Form awareness		
	5 Concept sentence		12	5 Morphemes	puppets	6
	6 Semantic / grammatical acceptability of sentences	puppet	12	6 Grammatical acceptability of sentences	puppets	6
			168			86

II The second part investigates the child's knowledge of some technical terms often used in reading instruction (a kind of meta-language belonging to the teacher's Reading Instruction Register). As a matter of fact, the teacher uses these terms from the very beginning, and it can be expected that many children will be confused. It is important for the teacher to have a clear notion of the children's knowledge in this respect. In the PI we use a test-booklet in which the child is asked, for example, to draw a circle round 'one word', or 'all the letters', given a series of printed stimuli.

III The last part of the PI tackles phoneme, word and form awareness. It is generally accepted that in order to learn to read, children have to bring their knowledge of the spoken language to bear upon the written language. This requires the ability to deal explicitly with the structural features of spoken language. It is, therefore, justified to suppose that some reflection on spoken language is necessary for the child to be able to discover the properties of spoken language that are central to the correspondences between the written and the spoken mode. The PI investigates metalinguistic awareness of the spoken language by means of six tasks, at three levels:

Phoneme Awareness

III1 Phoneme differentiation. In this task the child has to give the first phoneme of a given word. Before this the child is instructed on what is meant by 'first sound in a word'. Afterwards two toy frogs are introduced (a little one and an old king frog). The old king frog wants to play a game with the little one. The game consists in pronouncing the first sound in a given word. The children are asked to help the little frog, because it does not know how to perform this task.

III2 Rhymes. The children are first asked if they can give a word that sounds like 'roos' (rose). A few examples are generated. This procedure is repeated with the word 'beer' (bear). Then the child is acquainted with a hand-puppet called 'Jos' who likes words that sound like his name. 12 test words are presented in random order (six rhyme, six don't). For each word presented the child should answer the question 'Does...sound like Jos?'.

Word Awareness

III3 The concept 'word'. In this part we want to look into the child's conceptualisation of 'word'. Both function and content words are used as examples. 12 items are presented in random order. No feedback is given. For each item the child is asked: 'Is...a word?' Six words, three phonemes and three sentences are included.

III4 Long / short words. In this task we first present the child two words and tell which word is the longer one and which is the shorter. Then we present 12 pairs of words (in randomized order as to length). No feedback is given.

Form Awareness

III5 The concept 'sentence'. In this subtest we try to tackle what prereaders understand by 'sentence'. Again 12 items are presented (in random order: six sentences, three phonemes and three words). For each item the child is asked: 'Is ... a sentence?' No feedback is given.

III6 The semantic / grammatical acceptability of sentences. In this subtest we use a hand-puppet that is learning to talk properly, but whose speech is not yet correct. Sometimes it says the things correctly, but sometimes it says things the wrong way round. The child is asked to help to teach the puppet how to talk properly by judging whether the spoken sentences presented to him are right or wrong. A few examples of such sentences are: 'Tanden je poets' (teeth your brush) (syntactically incorrect), 'De jongen zwemt in de lucht' (the boy is swimming in the air) (semantically incorrect). 12 items are presented in random order. They include: 6 grammatically wrong sentences (but semantically correct), 3 semantically wrong (but syntactically correct), and 3 correct sentences. No feedback is given.

Scoring with the PI is easy: for each correctly solved item the child receives 1 point.

3. THE TRY-OUT AND FIRST INVESTIGATION WITH THE PI

A sample of 131 kindergarten children were screened just before they entered the first grade of elementary school. Because of absenteeism or organisational difficulties only 104 were administered the complete PI. We obtained reading achievement scores for 100 Ss (50 male, 50 female). Two of them were given by the teacher(s): schoolmarks for reading (after 4 months of instruction), and a score on a Likert-type scale (5-point scale ranging from very weak to very good reading competence). The third score was obtained by means of a standardized reading test (Enschedese Leeskaart 2) which was individually administered to the Ss. In this test the child is asked to read a story which is made up of 'regular' one-syllable words, printed without capitals. The score is computed from the correctly read words in one minute. The test was administered after half a year of formal instruction.

All four scores have been correlated, using the Bravais-Pearson formula. Table 2 presents the coefficients of correlation. All correlations are significant at the .01 level. There are no significant differences between boys and girls on the PI. Our initial hypothesis seems to be confirmed: children who begin formal reading instruction with a higher level of metalinguistic awareness achieve better after half a year of reading instruction than the children with low metalinguistic awareness. It should be noted, however, that these findings do not permit any statement on the possible causal influence of metalinguistic awareness on reading achievement. In a study which is in progress possible interfering factors (such as social background, school readiness, other aspects of language development, IQ) will be controlled. A more sophisticated and stronger statistical analysis of the results obtained by a much larger sample will be envisaged.

4. THE GTI

In the study in progress just mentioned, a shortered version of the PI is used. To that effect we determined, for each item of the PI, the index of difficulty (the proportion of Ss that gave a correct answer to the item), as well as the discrimination index (the correlation of each item with the subject's total score on the subtest in which the item figured). Both indexes were computed according to Davis. PI- items which were either too difficult (passed by fewer than 20 % of the Ss) or too easy (passed by more than 80 % of the Ss) were deleted. Non-discriminating items (discrimination-indexes below .35) were also eliminated or revised (see table 3).

The reliability of the PI was also determined. We used Gronbach's α (executed according to the SPSS-program). With dichotomous

Table 2: Bravais-Pearson coefficients of correlation (PI).

	I	II	III	TOT
Eschedese leeskaart	.36	.52	.54	.58
5- point scale	.36	.49	.68	.60
schoolmarks reading	.28	.44	.49	.51

(all significant on .01)

items α is equivalent to Kuder-Richardson's Formula 20 coefficient of reliability a widely used method for determining the reliability of a test. Coefficients of .80 (at least) or .90 (preferably) are desirable when predictions are to be made. As shown in table 3, two subtests and the total test obtained coefficients of reliability of .90 or above. Four others obtained coefficients of reliability of .80 or above. The most striking coefficient is in III3 (the concept 'word'). Ss were probably guessing when asked for their implicit criterion for 'word', or did not possess a reliable criterion to judge what was a word, what was not. PI- subtests III3 and III5 (the concept sentence), as well as III6 (the acceptability of sentences) have been reworked or completely changed. From subtests II (recognizing literacy behaviour), III2 (rhymes), and III4 (long/short words) items having low and/or unproductive indexes of difficulty were deleted. The new versions of III3, III5 and III6 are as follows:

III3 has been replaced by a task in which Ss are asked to indicate word boundaries. A sentence is spoken to the child; for each word he or she should lay down a cube. Six sentences are presented, differing in the length and types of words used. A justification for the inclusion of this test can be found in the often reported correlation between this sort of activity and reading achievement.
III5 has been changed into a task in which the child's ability to judge the morphemic endings of words is tested. The child is acquainted with two puppets: one, called Rankie-Pankie (a character played by the experimenter), is still learning how to talk properly; the other, called Mies (a character to be played by the child) is helping. The child is told that now and then Rankie-Pankie says words the wrong way and that Mies' job consists of helping him to pronounce them appropriately. An example is given. For each item a picture is shown together with a verbal context for the test word. For example: Rankie-Pankie is shown a large dog and a small dog. At the same time the experimenter asks: 'Een kleine hond is een...?' ('a small dog is called a...?'). The correct answer would be: 'hondje' ('doggie'). If Rankie-Pankie gives the wrong answer, Mies (the child) should correct him. If the child answers: 'wrong', it is asked to say it correctly in order to help Rankie-Pankie. There are six items. Morphemic endings tested include: third person simple present tense, plural, diminutive. For each of these a correct and an incorrect item is given (at random).
III6 concentrates on word order. For this form judgment task the two puppets are used again. Mies (the child) has to judge of the sentences uttered by Rankie-Pankie (the experimenter) sound good or funny. There are six items: three correct, three incorrect sentences (at random).

The scoring for the GTI is the same as for the PI: for each correctly solved item the child receives 1 point (total score: 86).

Table 3:Reliability (PI)

	I	I	TOT I	II	III	III	III	III	III	III	TOT III	TOT
Cronbach α	.72	.90	.89	.86	.89	.75	-.21	.75	.65	.48	.83	.92
$\sqrt{x}$	1.76	2.10	2.83	2.47	1.24	1.17	1.66	1.39	1.54	1.39	3.58	5.20
k	32	28	60	36	12	12	12	12	12	12	72	168
N	131	131	131	131	104	131	131	131	131	104	104	104

Cronbach's α : reliability coefficient
$\sqrt{x}$: standard error
k : number of items
N : Ss

5. CONCLUSION

The motivation for the study reported in this paper can be found in our belief that metalinguistic awareness plays an important role in learning to read, and probably in other learning tasks as well. The preliminary findings are encouraging: the predictive validity of the PI c.q. the GTL is satisfactory. Moreover we think there are solid psycholinguistic grounds for ascribing construct validity to the operationalisation for metalinguistic awareness that we are using. If in the near future the GTL proves to be an interesting psychodiagnostic device, norms will be worked out. But, for the time being, we regard it primarily as a promising research instrument which could be useful in measuring the factor 'metalinguistic awareness' in studies that will investigate the relationship between factors that are supposed to have something to do with the acquisition of reading, and reading achievement. Whatever its use, however, we hope that it will have a positive impact on psychoeducational and instructional practice.

ACKNOWLEDGEMENTS

We would like to thank P. Vereecke and his students of the section Functionele Grafiek of the St.-Lucas Higher Institute Gent, for their skilful help in designing the test material. M. Bijnens and Y. Cops, both final year students at the Katholieke Vlaamse Hogeschool Antwerpen, had an essential share in the administration of the tests. We appreciated very much the support of the Laboratorium voor Toegepaste Psychologie (Metrische Psychodiagnostiek) RUG, in the computer processing of our data. Part of this research was performed with the help of a grant to the senior author from the Research Fund of the Rijksuniversiteit Gent.

REFERENCES

Spoelders, M., 1982, Psychopedagogische verkenning van (leren) lezen via case-study. In: Handelingen van het 2de internationaal IRA-symposium, Belgische sectie over leesstoornissen (Antwerpen 1981).

Spoelders, M., and van Damme L., 1982, Early metalinguistic awareness and reading. In: Scientia Paedagogica Experimentallis 19,1,117 - 129.

Spoelders, M., and van Damme L.; 1983, En de Leeuw? Hij las voort. Het leesonderwijs in Vlaanderen. In: En nu over jeugdliteratuur 10, 6, 220 - 222.

van Damme, L., 1984, Geïndividualiseerd Taalbewustzijnsonderzoek bij het Leesbegin. GTL (RUG: Laboratorium voor Pedagogiek, experimentele versie).

van Damme, L. and Spoelders, M., 1983, Luisteren en blijven luisteren. In: MOER 6, 8 - 13.

van Damme L. and Spoelders, M., 1984, Metalinguistic awareness, and reading (Paper read at the 7th AILA Congress, Brussel 1984).

van Damme, L. and Spoelders, M., 1985, Reflections on a metalinguistic awareness test for prereaders. In: M. Spoelders, Fr. Van Besien, Fr. Lowenthal, F. Vandamme, eds., Language acquisition of learning. Essays in educational pragmatics-2 (Leuven, Acco and Gent, c of c), 67-78.

GOALS, METHOD, AND THEORY IN LANGUAGE INSTRUCTION

H. Stephen Straight

State University of New York at Binghamton

Binghamton, NY 13901 U.S.A.

The present paper addresses three very different questions about language instruction. First, WHY is there language instruction? A review of the variety of reasons why people might have a use for learning a language leads to the conclusion that a very great variety of methods of language instruction should be employed.

Second, however, on the assumption that the learner's goal has been identified to be the most ambitious possible goal, namely the achievement of native-like fluency in all areas of linguistic performance, HOW can this goal be most effectively attained? Research on this topic suggests, paradoxically, that the most effective teaching methodology for the achievement of the full range of skills is one in which the sole focus of the teacher and of the learner is on comprehension skills, both listening and reading, while production skills are left to develop more or less on their own and at the discretion of the individual language learner.

Third, in attempting to resolve the paradox just noted, WHAT does the language learner learn in the process of language instruction? In this case, it turns out that the wildly counter-intuitive finding that the building of comprehension skills leads to the establishment of the major part of production skills follows naturally from a proper understanding of the relationship between the four domains of language processing--receiving, interpreting, formulating, and executing--which are involved in the comprehension or production of any linguistic act.

INTRUMENTAL VERSUS INTELLECTIVE GOALS IN LANGUAGE LEARNING

Language teachers generally assume that the proper goal of language instruction is to develop the full range of linguistic skills of the imagined typical fluent native speaker-listener.

Recently this typical but very ambitious goal has been modified in recognition of the fact that many learners have much more limited "specific purposes" for which they intend to use the language (Wilkins 1974, 1978; Munby 1978; Mackay and Mountford, eds., 1978; Mackay and Palmer, eds. 1981). This has led to the creation of highly specialized teaching materials, emphasizing specific content areas, such as hotel-keeping, import/export transactions, travel planning, etc. It must be noted, though, that the desired endpoint of the instruction in these cases is still the achievement of fluent communicative skill, but in only a limited arena of linguistic interaction.

The common belief of both teachers and learners that communicative fluency, however full or partial it may be, should be the predominant goal of language instruction neglects a number of facts. For one thing, there are many cases in which the goals of language learners, or the predisposition of language teachers clearly go beyond mere communicative fluency to include such skills as translation, simultaneous interpretation, creative writing, and literary criticism. However, it is clear that such additional goals are just that: additional. The primary goal upon which they are added is communicative fluency.

More important, then, is the failure to recognize that some language learners clearly have needs that fall far short of even this primary goal. For example, many learners have no more loftier goal in mind than simply to be able to get along for a week of travel overseas. They may, in fact, not even want to limit themselves to learning just one language. If they expect to be in several different language communities, they might better learn a number of languages in the roughest of "pidginized" versions, sufficient to satisfy their various but rather limited needs. Clear signs of their marginal linguistic ability might be intentionally maintained in their speech in order to keep fluent speakers from jumping to false conclusions about the extent of their communicative capacities. In many instances, the learners would clearly be better off learning to speak in a manner that would indicate immediately that any response to their simple inquiries or requests must also be clear and simple.

However, even the example just given still fits clearly in the category of what I would prefer to call the instrumental goals of language learners. Differences along these lines clearly deserve to be recognized by language teachers in their design of wide variety of different "specific-purpose" instructional materials and teaching techniques. But there are also various non-instrumental goals that, although they are very commonly pursued in conjunction with language instruction, have seldom been recognized as having a bearing on the way in which languages might be taught.

I will call such non-insrumental goals intellective. By this I mean to refer to two types of language knowledge. First, the conscious structural knowledge that linguists typically acquire in the course of their study of languages. And second, the conscious cultural knowledge that language teachers typically emphasize when they argue for the place of foreign language learning in general education curricula. The goal of such study, both for linguists and for literati, is said to be to inform the learners concerning the nature and extent of either the linguistic or the cultural differences that differentiate them from the members of a linguistic group other than their own. In the case of a purely intellective program of instruction, no expectation of the achievement of communicative fluency need be made. The typical instrumental language class should focus upon the establishment of the subconscious processes of language perception and production. In contrast to this, an intellective language class focusses upon the accumulation of conscious items of fact and theory concerning the language and the people who use it.

In an intellective class, emphasis can be placed from the very start on analysis of linguistic structures or on esoteric cultural artifacts. The teacher can present descriptions of the sound system, grammar, and semantic structures of the language without any necessity to get the learners to memorize and practice any rules, words, or phrases. And works of art or historical accounts of the people who use the language being taught can be discussed and analyzed in the learners' native language, perhaps by means of side-by-side translations. The hope is that after a year or two of such study the language learners, despite their obvious lack of instrumental skills, will nevertheless have acquired a deep appreciation of the ways in which Language (with a capital L) both unites and differentiates human groups.

Even if all of these intellective goals are seldom achieved, it is certain that such liberalizing and informing benefits of language instruction are often paramount in the minds of the educators who propose language requirements and of the pupils and parents who accept them. However, there is considerable reason to doubt that such non-instumental goals are in fact well served by the usual foreign language class.

Intellective goals of the linguistic sort would be better served in a class devoted solely to the description and analysis of the structure of a given language. Indeed, a fully comparative-linguistic approach should succeed in giving learners a far greater appreciation of the complexity, diversity, and universality of linguistic structures than mere parroting and memorizing could ever hope to achieve. Similarly, intellective goals of the literary or cultural sort would probably be served by instruction emphasizing informative accounts of the ways of life of the groups who use a particular language. Such instruction could presumably employ translations and other materials readily accessible to the student, with or without any prior knowledge of the language of those groups.

What is clearly problematic about most language instruction, then, is that the above linguistic and non-linguistic intellective goals are usually sought simultaneously with various instrumental or quasi-instrumental goals. This mixing of goals can lead to the expenditure of enormous amounts of time and effort in trying to achieve marginal communicative skills, to the detriment of the intellective goals. Moreover, the addition of such broad intellective goals to those of a solely instrumental instructional program is not a good idea either. Rapid attainment of particular communicative skills is made more difficult by loading learners down with linguistic and cultural information that will not be of any value in the particular communicative situations to which the instrumental instruction is directed. Even if full cultural knowledge is also desired, the rapid achievement of comprehension fluency will give the learner early access to literature history, and other primary sources in the language, sources that cannot be approached in the texts that are typically employed in beginning language classes.

I now can state the conclusion of this first part of my paper: The possible goals of language instruction vary tremendously along both instrumental and intellective dimensions in a way that is not reflected in the observed variety of instructional materials and methodologies. The notion that there is some one best way to teach languages is thus highly dubious in the absence of a clear specification of the goals of the instruction.

For the remainder of this paper, however, I will discuss the methods and theoretical basis for language instruction exclusively in the context of the most usual sort of instrumental goals. That is, I will assume that the "why" of language instruction is usually best answered as "to achieve the full range of instrumental language skills". To the extent that this is not the goal, the rest of what I have to say may have little bearing on the "how" of language instruction. For present discussion, then, I will assume that the learner is one who has good reason to want to achieve full fluency in the language and that the teacher is willing to focus upon this goal to the exclusion of intellective goals.

I will further assume in what follows that the learner and the teacher are agreed that full rather than partial mastery of communicative skills is the desired outcome of their joint efforts. How one might best go about designing a program of language instruction to achieve lesser degrees of fluency (for touristic, employment, or other "specific purposes") constitutes a separate research question, one that I have not explored.

COMPREHENSION VERSUS PRODUCTION IN INSTRUMENTAL LANGUAGE INSTRUCTION

A growing and diverse body of empirical research, some of it very recent, indicates that the best way to learn a language is to learn how to comprehend it fluently, and that everything else will follow, if not automatically, at least far more easily and effectively. Early suggestions for an emphasis on comprehension training were made in the '50's and '60's (Scherer, 1950, 1952; Belasco, 1967; Asher, 1969), but it was in the '70's that a fullfledged "comprehension approach" to language instruction emerged (Postovsky, 1974; Winitz & Reeds, 1973, 1975; Postovsky, 1974; Davies, 1976). In the early "80's this approach has come to occupy a prominent position in the field (Davies 1980; Nord, 1980; Gary & Gary, 1980, 1981; Krashen, 1981; Terrell, 1982).

Two recent book-length treatments of the comprehension approach deserve special mention: The first is a book edited by Harris Winitz (1981), containing articles by many different authors who have all concluded, each in a different way and on the basis of different kinds of evidence, that comprehension training ougt to precede production training, perhaps by as much as one full year of full-time instruction. The second is a book written by Robbins Burling (1982), a professor of linguistics and anthropology at The University of Michigan who has been working for years on an experimental program in the teaching of French in which reading and listening are taught to college-age learners to the virtual exclusion of production skills (Burling 1978).

A prime characteristic of the comprehension approach is that it avoids the teaching of explicit linguistic knowledge (Krashen 1982): "Learning" conscious facts and rules impedes the "acquisition" of the subconscious psychological processes of language performance that will enable the learner to make effective use of the language as a means of communication. Conscious attentiveness to one's own productions or to the productions of others gets in the way of fluent communicative exchange. Burling (1982) even goes so far as to say that the conscious precepts of "grammatical usage" are not useful in the process of recognizing and correcting errors: The most effective "monitor" is not conscious grammatical rule-following but rather the intuitive trial-and-error process of noticing that something doesn't "sound right" and then arriving at something that does.

Similarly, the best learning situation is one that recreates the supportive and relaxed atmosphere of the native language learner. Blair (1982) reprints a number of the more radical proposals along these lines, which are known under such names as "Community Language Learning" (Curran), "Suggestopedia" (Lozanov), and "The Silent Way" (Gattegno). All of these methods show that many things can be done to enhance the learners' emotional receptiveness. Examples include playing music, using lots of concrete objects, pictures, and amusing examples, and presenting topics that will be of immediate interest to the individual learners. Attitudes toward the language and its speakers and affective reactions to the learning situation influence the learners' performance greatly.

A feature common to all of the characteristics mentioned so far is that they serve to establish a format in which the learner is most likely to be attentive in the learning situation. Anything that might serve to distract or annoy the learner is to be avoided. Without any doubt, one of the greatest sources of annoyance and distraction for language learners is the frustration and effort they experience in trying to express themselves in a new language. Most learners, however, agree with their teachers that this annoyance and frustration is simply the painful price one must pay to learn a new language. Happily, advocates of the comprehension approach have concluded that the best language instruction method is one in which learners are not required to speak or otherwise produce in the language being learned.

That is, the best method of teaching a language focusses solely on the development of comprehension skills, including both listening and reading. Learners are required to show their developing comprehension ability in any of a number of ways, both non-verbally and in their native language, but they are not required or even especially encouraged to try to speak or write in the language being learned. And whenever the learner does begin to produce things in the new language, the teacher is not to correct errors. Instead, the teacher is to ensure that the learner receives comprehensible and well-formed input, regardless of the well-formedness of the learner's output.

Harris Winitz has provided a useful brief description of the key methods associated with the comprehension approach, minus the affective, motivational factors which have more to do with the setting and the background characteristics of the learners than with the design of instructional materials (1981: xiii):

> Listening comprehension is stressed. Students are given an opportunity to acquire the grammar of a second language by

> acquiring a fundamental understanding of the language. Initially the student hears sentences in the second language for which the meaning is clearly indicated through actions or pictures. All sentences are spoken by native speakers. Lexical items and grammatical units are carefully programmed to maximize implicit learning. The umbrella term for this instructional system is the comprehension approach. It differs significantly from other methods in the value placed on speaking exercises. Production drills and preplanned dialogues or artificially generated conversations are not used. Conversation is not discouraged. It simply is not taught. The belief is held that conversational fluency will develop as the result of learning to understand a language.

Pattern drills and pronunciation practice are replaced by a seemingly super-permissive acceptance of any output the learner may produce, as long as it is intelligible in any way to the teacher. In fact, the main thing appears to be to keep the learner actively and communicatively involved in the teaching situation.

The instructional experiments so far (see Winitz, ed., 1981; and Burling, 1982) have shown that the success of the comprehension approach is not entirely uniform. Instead, it seems that such success is directly proportional to the experimenters' success in devising instructional materials that (1) are not too difficult for the learners to understand but (2) are at the same time of compelling interest to them.

On the first point, Heidi Dulay, Marina Burt, and Stephen Krashen (1982) note that "The specification of principles governing learning complexity is probably one of the most important areas of theoretical research that remains to be undertaken" (58). Despite this continuing uncertainty as to exactly what the best order of presentation is, experimenters have devised any number of ways to increase the degree of difficulty of presented material in a graded manner. These have ranged from merely intuitive simplification of the teacher's speech in the classroom and in prepared materials to the ordered introduction of specific complexities of grammar and vocabulary (as gauged by order of mastery in child or adult language acquisition, see Dulay et al. 1982) to the actual mixing of vocabulary and structures from the language being taught and the language already known to the listeners (see Burling 1978).

On the second point-- the maximization of communicative involvement--experimenters have been especially creative. The "Total Physical Response" method of James Asher (1977), for example, emphasizes the acting out of simple commands by learners from the very first day of class, with increasing complexity of non-verbal response thereafter. The late Valerian Postovsky (1981) stressed the "synchronized presentation of language material and pictorial events" (1976) followed by a multiple-choice selection among an array of pictures on the basis of a recorded utterance. Another way to capture learner interest is to present information relevant to the solution of simple problems exclusively in the language being learned; Problem-solving will thus be dependent upon accurate comprehension (see Winitz & Reeds, 1975). A final and even more general strategy for the teacher to employ in selecting and ordering material for presentation is to present topics that make direct reference to issues in which the learners have a real personal interest or significant background knowledge (see various authors represented in Blair, ed., 1982).

The empirical evidence for the effectiveness of the comprehension approach as described above is presented and discussed in the works I have cited. I will not rehearse it here. Nor will I discuss the issue of when, how, and why to introduce production tasks, though I am prepared to suggest that some memorized phrases could be introduced at an early point to serve as models for learners to refer to when they choose to begin speaking. Instead, I would like to turn now to the final point of this paper: Why should it be that an approach that avoids explicit instruction in production skills manages to foster both production and comprehension skills?

LANGUAGE INSTRUCTION IN THE CONTEXT OF A THEORY OF LANGUAGE PROCESSES AND THEIR INTERACTION

The obstacles to an acceptance of a comprehension-centered approach to language instruction are numerous. One of the greatest is that of the common-sense notion of what it means to "know" a language: If I tell you that I know French but can't speak a word of it, you are likely to say that, well, yes, maybe I understand it, but to know a language is to be able to speak it. This bias toward production is profound. Students in a comprehension-centered language class feel cheated because they are unable to say or write anything in the language they are learning, even if they are comprehending far more than their parrot-like counterparts in the typical audio-lingual class.

There is far more working against the best method than just common sense, though. Scholars and teaching methodologists have long believed that knowledge of a language depends critically upon knowledge of how to produce utterances. An everyday corollary of this belief is the old maxim that you don't really "know" a word until you have used it yourself. Unfortunately, this bit of conventional wisdom is contradicted by the universal superiority of passive vocabulary over active vocabulary. A second, more esoteric variant of the alleged primacy of production is the claim that "analysis" of speech input depends upon the tacit "synthesis" of it. Naturally, this more esoteric hypothesis must be defended or disputed with similarly esoteric evidence, but it is nonetheless contradicted (see Straight 1976, 1980).

A third variant of the production-first argument is the now largely discredited position of the radical behaviorists, who seemed to insist that the only kind of learning that occurs is learning how to move one's muscles. The audio-lingual methodology with wich so much language instruction is still saddled grew out of the notion that to learn a language is to mimic the articulatory output of a native speaker while having that output monitored by a reward-administering teacher. It was thus fascinating to see Stanley M, Sapon, who is radical behaviorist of considerable stature and tenacity, come around to the recognition that

> From the point of view of linguistic analysis, <u>responses</u> are made only by <u>speakers</u>. This leaves us in the odd position of describing, analyzing, and predicting the performance of agents in a communication system on the basis of the <u>formal</u> properties of <u>one-half of the system</u>! (1971:80)

So much for the behaviorist bias toward production!

On the positive side, advocates of the comprehension approach offer a large number of explanations for the success of their approach. The simplest of these is that it is easier to teach learners to understand spoken and written language than it is to teach them to produce it. Learners thus acquire a greater amount of functional skill after a given amount of time and effort spent in acquiring comprehension skills than they have after that same amount of "mixed" training (see Scherer, 1952). This argument for effectiveness is sometimes further defended on the grounds that the motivational and motor-coordination obstacles to speaking are too great to overcome in the early phases of language learning (Dulay et al. 1982: 4). The success of the comprehension approach is, by this argument, simply a result of the dangers of early production training. No mention is made of the rather mysterious fact that production skills appear to be developed in the comprehension approach.

More theoretical arguments are also advanced, to be sure. One of the more common of these is neuroanatomical, and it takes two basic forms. First, it is noted that early production training may result from the activation of a mimicry routine involving uncomprehending repetition of input via the direct activation of the motor speech area from the lowest level of receptive processing in the auditory cortex. That is, early (as well as later) speech output may be produced on the basis of less than adequate understanding of what is being produced (Diller 1981). Second, it is noted that comprehension skills may be represented in both the left and the right hemispheres of the cortex rather than in only the left hemisphere. This would mean that learning might take place faster, and perhaps with a greater involvement of the entire psyche of the learner, if instruction emphasized bilaterally represented skills (Asher, 1981). Again, of course, the production skills developed via the comprehension approach are not mentioned.

Other arguments are more theoretical still. Putting comprehension first in instruction is seen to follow the "natural" example of first-language acquisition: Children acquire comprehension skills before they acquire production skills (Asher 1972). This primacy of comprehension is also seen to follow from the "logical" priority of comprehension: Production of intelligible output is, for the most part, dependent upon the copying of some comprehended input; neologisms are neither prevalent nor long-lived.

I have no quarrel with any of the above explanations of the effectiveness of the comprehension approach. In fact, I am willing to believe that they all contribute to our developing theory of the nature of language and its acquisition. There is, however, one additional line of argument that I find highly dubious, although it is challenged by no one and espoused by many. It is the sole argument advanced in explanation of the production-developing aspects of the comprehension approach. Let me turn to this argument now, as my objections to it constitute the primary point of this third and final part of my paper.

In the lead article of the 1981 volume of I.R.A.L. (the International Review of Applied Linguistics), Judith Olmsted Gary and Norman Gary made perhaps the most detailed theoretical presentation of rationale for "a much greater emphasis on listening comprehension for language instruction". The relationship they delineate between comprehension and production is historically typical, though the way they phrase it is both clearer and more current than any I have found. Let me quote at some length from their article (1981:2-3):

> ... In terms of linguistic competence, listening and speaking require the same kind of language knowledge, not two different kinds. The same set of rules ... are used in language comprehension as in language production. This is illustrated graphically in Figure 1.
> ...The major differences between the encoding and decoding tasks--aside from the motor movements required for speech or writing--have to do with processing: information retrieval, short and long-term memory processes, control of speech of processing, etc. Also there are different strategies required in processing for comprehension and production. In production the speaker controls the structures and vocabulary used; thus a foreign language learner can use

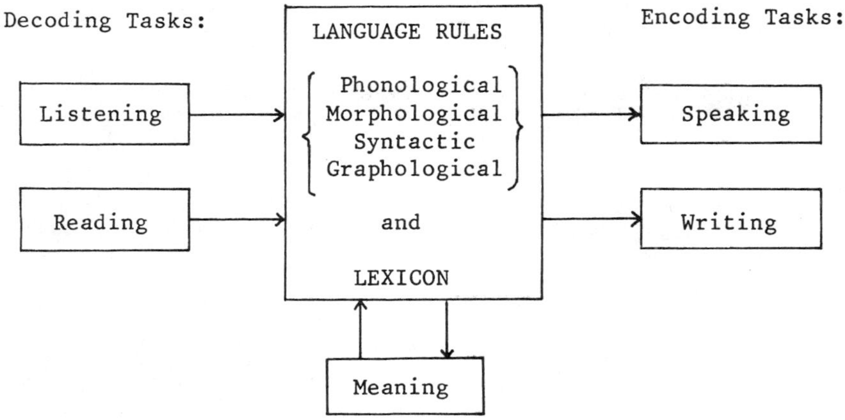

Fig. 1.

various paraphrastic devices when he is producing language, avoiding problematic lexical and grammatical structures. But when he is comprehending, he does not have this control; he must deal with an essentially open grammatical and lexical system (Ingram, et al., 1975). But these differences aside, the underlying system of the language is the same for both comprehension and production.

At first glance, the observation that speakers can limit the processing difficulty of their output while listeners cannot similarly limit the processing difficulty of input appears to argue against the claim that listening is "easier" than speaking. This contradiction is resolved as follows (Gary & Gary, 1981:3):

> ... Speech requires linguistically more complex tasks than comprehension. Comprehension--at least at all but the most advanced levels--allows many linguistic signals to be ignored: redundant grammatical and semantic functions such as concord, definite/indefinite distinctions, singular/plural distinctions, etc., can very often be ignored without seriously distorting the message being comprehended.

The Gary's argument for the primacy of comprehension in language instruction is, thus, heavily based on a recognition of the practical difficulties faced by speakers, plus some recognition of deeper processing differences between the various linguistic tasks. This is little more than a repetition of the observations already presented in the present paper, with which I have no disagreement.

What I do find troubling is the theoretical framework in which they have couched their explanation of the rapid and unforced emergence of production skills in later stages of language learning even though the instruction gives these skills no attention. In attempting to account for this phenomenon, they fall back on the standard notion that language knowledge is neutral as regards component skills of comprehension and production (1981: 3):

> ... Since the same set of rules underlies both speaking and listening, any instructional methodology which teaches the system will of necessity have effects on the total language competence.

Some of you have heard me present arguments against this line of argument on previous occasions. In fact, those of you who were present at the last conference in the present series heard an entire session concerning the irreducibility of the contrast between production and recognition (see Straight, 1982 and the other papers in the section of the volume in which it appears, pp. 141-201). For present purposes, it is enough to say that there is no empirical evidence to support the claim that language knowledge exists separately from knowledge of how, on the one hand, to process linguistic input and how, on the other hand, to produce linguistic output. In fact, there is no theory of such performance-neutral knowledge that can be used to test the claim that it can be used, even by an "ideal" speaker-listener, to accomplish anything whatsoever. Models of competence, and the theories from which they derive, remain as abstract mathematical formalisms that are, at best, only heuristically valuable to those who are seeking to develop a theory of language knowledge (see Matthews, 1979).

Fortunately, there is an alternative to this failed line of argument, and the Garys themselves provide an entry into it (3):

> ... To speak communicatively, one must comprehend incoming messages and then organize appropriate responses for production.

I believe that they are almost exactly on target here: Speaking depends upon (a) comprehending input and (b) organizing responses for output. The rest of this paper will consist of an attempt to put this observation into the context of a theory of how the various processes of understanding and saying are related to each other. The figure below contains a diagrammatic summary of the overall theory. What remains to be done is to show how the findings of the comprehension approach are in accord with the predictions of the theory.

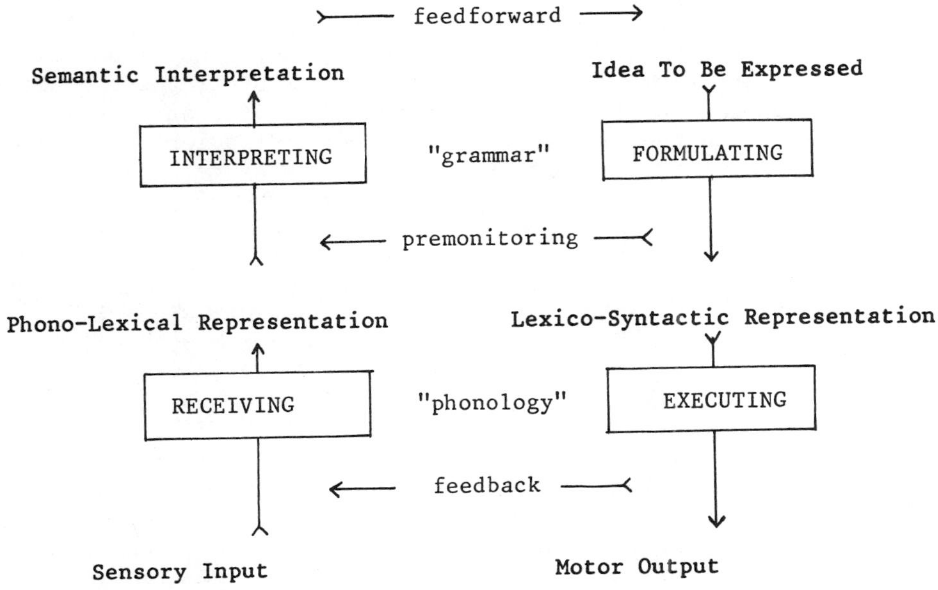

Fig. 2. Diagram of Language Processes and Their Interaction

The theory summarized diagrammatically above finds no place for any sort of processing-neutral language knowledge: The labels "phonology" and "grammar" that appear in the diagram are useful ways of characterizing the auditory and articulatory versus the higher-order perceptual and cognitive processes that operate upon linguistic input and organize linguistic output regardless of the physical channel being employed. Following Gary and Gary, then, "graphology" could appear in place of "phonology" in order to have the diagram serve to account for reading and writing. All the other labels would remain as they are, however, except that--by implication--the Sensory Input would be visual and the Motor Output would be manual, and concomitant changes would be implied for the nature of the RECEIVING and EXECUTING processing components and possibly also in some aspects of the higher-order levels of representation and processing. In any case, though, the absolute separation of comprehension and production processes is retained.

On the other hand, these processes, however separate they are, do interact. At the lowest level, the motor output of the language producer typically results in sensory input to that same person as self-perceiver: This kind of reflexive processing is labelled "feedback" in the diagram. Higher up, but still at a rather concrete level of representation, where words and phrases are consciously available, language users are, of course, able to engage in anticipatory self-comprehension of linguistic structures not yet executed as motor output. This internal feedback processing is labelled "premonitoring".

Above the level of what is called Phono-Lexical and Lexico-Syntactic Representation are the processing components labelled interpreting and formulating. The input and output to these components is decidedly non-conscious. Speakers are aware of what they are about to say, and they may of course be aware of a large number of non-linguistic impressions that are felt to be related to what they are about to say, but they are no more aware of the meaning of it than they are of the meaning of what someone else says before it is said. The process of premonitoring is thus a process of discovery, and even at that the only thing that is discovered is that what one is about to say leads one to think of saying other things, such as "No, that's not what I mean," or "Yes, I did say that well". (see Straight 1978 for a fuller discussion of the role of consciousness in all of this).

I will not elaborate upon this model of language processes here (though I have done so previously and plan to do so further in the future and would be happy to do so in the discussion session later, if you like). Instead, I will show briefly how the model meshes with the evidence for the primacy of comprehension skills in language acquisition.

Notice first that according to the model "feedback" and "premonitoring" are simply reflexive channels of connection between executed or formulated structures. They constitute labels for the fact that what you are thinking of saying or what you have already said serves as input directly into the receiving and interpreting mechanisms. What needs to be added to this point is simply that there is considerable evidence from studies of flawed auditory feedback and dysphasic symptoms that this kind of internal processing is essential to the production of well-formed linguistic structures (see Straight, 1976 for expansion of these observations). When people, for whatever reason, are prevented from hearing their speech output, that output becomes distorted considerably. Similarly, when a brain-damaged patient exhibits an inability to comprehend linguistic input, that patient's linguistic output also shows considerable semantic distortion, even though it may be fluent from a phonetic point of view.

Accurate production of linguistic output, whether actual or merely anticipated, this depends crucially on properly functioning feedback and premonitoring mechanisms. It should be clear without my saying it that the correct functioning of these mechanisms depends in turn upon the presence and correct function of comprehension abilities, whether those that apply to sensory input (receiving) or those that apply to recognized phono-lexical phrases (interpreting). It thus follows automatically that production abilities develop most accurately and fully only after comprehension abilities are well established. Thus the model of language processes presented here predicts the success of a comprehension approach to language instruction.

My story is nearly complete. All that remains is to explain why specific training in production, other than the "training" that comes automatically from expressing one's ideas and seeing whether the way one has done it is in fact intelligible and acceptable, does not appear to be necessary in order for language learners to develop their production skills.

The solution to this puzzle may, I believe, come from a careful consideration of the role of "feedforward". Feedforward is the series of processes that are applied to the results of interpretive processing to arrive at the various associations, implications, and possible supportive or conflicting responses that might be given to those interpretations. It is also the basis for the kind of anticipation of what a speaker is going to say next that is so important in the development of fluent comprehension skills (see Straights 1982).

James Nord (1981) has provided perhaps the best previous discussion of the role of what he calls the "Anticipatory Feed Forward Phase" in the development of listening fluency. His description of the techniques that have been found useful in developing such anticipatory skills includes one (described in Mackey 1971 and Landis 1978) that is particularly interesting given what has been said here about the comprehension approach to instruction (90-91):

> The teacher speaks only in the target language, but the student can speak in whichever language he or she feels comfortable. ... If the teacher is a native or near native speaker of French, students will always be hearing native French. They will not be spending half of the classtime listening to the poor pronunciation and grammatical misconceptions of their classmates, which is usually the case in classes in which speaking the target language is emphasized. This type of bilingual conversation can also be used with language tutors who have a poor speaking ability in English. If the tutors can comprehend enough English to maintain discussion, they can carry on a bilingual conversation. This way both the student and the tutor can increase their listening fluency in the new language while speaking in their own native tongue. It makes it easier on both parties, it develops the language competence through listening, and it is more interesting to both tutor and student because they can express themselves much better in their own tongue. Most importantly, they can discuss topics they are interested in, control the conversation more, and thus anticipate responses from other speakers.

What this means is that until such time as one has developed relatively accurate and complete understanding of what one is receiving, it will not be possible to arrive at appropriate Ideas To Be Expressed, not even in one's first language, to say nothing of a language one is just learning. However, once one does begin to have such ideas, the possibility for FORMULATING linguistic (and other) responses emerges naturally in the course of anticipating other speakers' utterances in ongoing listening. At that point, the attempts one makes to begin creating meaningful output modelled upon the linguistic input one has been getting occur in a fully informed and sensitive manner, without artificial, meaningless parroting.

Furthermore, because the processes of RECEIVING and INTERPRETING are fully developed, the learner's output requires no special intervention on the part of the teacher other than to respond as a normal participant in linguistic interaction with the learner who is trying to speak. Overt correction is unnecessary: Between the learner's own self-comprehension and the evidence of failed production that is implicit in the teacher's (or other hearer's) misunderstanding or incomprehension, the learner needs nothing more to develop production abilities. In fact, following Nord (1981) again, it appears that if learners are allowed to develop their comprehension fluency first, instead of developing bad habits through premature attempts to produce, they will naturally reach a "Discrepancy Feedback Phase" in which they "can detect discrepancies and errors because they sound wrong" (93). He concludes that under such circumstances "speaking skills should come very rapidly without specific speaking instruction because the trained listening ear serves as a self-monitor" (98).

Full productive linguistic ability develops therefore as a virtually automatic by-product of the development of comprehension processes through exposure to high-quality input presented in comprehensible contexts. The paradox engendered by the success of the comprehension approach to language instruction disappears in a theory of language processes that makes a clear separation among the constituent abilities that constitute language knowledge while emphasizing the ways in which these abilities interact in the ongoing comprehension, production, and acquisition of language.

ACKNOWLEDGEMENTS

This paper is a revision, expansion, and partial compilation of three papers presented at various times over the past several years: "Intellective versus Instrumental Goals in Language Instruction", Intensive Language Training Symposium, SUNY at Binghamton, 19 april 1975; "Foreign Language Instruction: Some Psycholinguistic Observations", Invited Lecture, University of Bucharest, Romania, 6 March 1980, and Anglo-America Week Programs, Constanza and Sibiu, Romania, 17 & 24 April 1980; "The Comprehension-Production Dialectic in Second Language Acquisition", Teachers College, Columbia University, 15 March 1981.

Let me acknowledge here my gratitude to the organizers of the programs mentioned above for inviting me to participate, and to the United States Mutual Educational and Cultural Exchange (Fulbright-Hays) Program and the New York State Council on Linguistics for financial support of the presentations in Romania and New York City, respectively. I also am grateful to the State University of New York at Binghamton for its support of my attendance at the above meetings and also at the present conference. Finally, let me thank the organizers of the present conference for their invitation and for their additional support. (Belgium has proven to be a most gracious and generous host).

I must also heatily acknowledge the (largely anonymous) debt I owe to the audiences on all of these public occasions: Each of them contributed in one way or another to the shape of the present argument. In fact, the title of the present paper underwent change as a result of hearing the other papers presented at the present conference! Composition is a neverending process.

REFERENCES

Asher, James, J., 1969, The total physical response approach to second language learning. Modern Language Journal, 53: 3-7.

Asher, James, J., 1972, Children's first language as a model for second language learning. Modern Language Journal, 56, 133-139.

Asher, James, J., 1977, Learning another language through actions: the complete teacher's guidebook. Sky Oak Productions, Los Gatos, Calif. Pages 36-52 reprinted as "The total physical response approach", in Blair, ed., 1982: 54-66.

Asher, James, J., 1981, Comprehension training: the evidence from laboratory and classroom studies. In Winitz, ed., 1981; 187-222.

Belasco, Simon, 1967, The plateau; or the case for comprehension: the "concept" approach. Modern Language Journal, 51: 82-88.

Blair, Robert, W., ed., 1982, Innovative approaches to language teaching. Newbury House, Rowley, Mass.

Burling, Robbins, 1978, An introductory course in reading French. Language Learning 28: 105-128.

Burling, Robbins, 1982, Sounding right: an introduction to comprehension-based language instruction. Newbury House, Rowley, Mass.

Davies, Norman, F., 1976, Receptive versus productive skills in foreign language learning. Modern Language Journal 60: 440-443.

Davies, Norman, F., 1980, Putting receptive skills first; an experiment in sequencing. Canadian Modern Language Review 36: 461-467.

Diller, Karl, C., 1981, Neurolinguistic clues to the essentials of a good language teaching methodology: comprehension, problem solving, and meaningful practice. In Winitz, ed., 1981: 141-153.

Dulay, Heidi, Burt Marina & Krashen Stephen, 1982, Language two. Oxford University Press, New York.

Gary, Judith Olmsted & Gary Norman, 1980, Comprehension-oriented foreign language instruction--an overview. The Linguistic Reporter 23(3): 4-5.

Gary, Judith Olmsted & Gary Norman, 1981, talking may be dangerous to your linguistic health! The case for a much greater emphasis on listening comprehension in foreign language instruction. I.R.A.L. 19: 1-14.

Ingram, Frank, Nord, James, R. & Dragt, D., 1975, A program for listening comprehension. Slavic and East European Journal 19: 1-10 (Cited in Gary & Gary 1981:3)

Krashen, Stephen D., 1981, The input hypothesis. In James D. Alatis, ed., 1981, Georgetown University Round Table on Language and Linguistics, 1980, Georgetown University Press, Washington, D.C. pp. 168-180.

Krashen, Stephen D., 1982, Theory versus practice in language training. In Blair, ed., 1982: 15-30.

Landis, G.B., 1978, Eureka! A surefire second language curriculum. System 6: 148-157 (Cited in Nord 1981).

Mackay, Ronald & Mountford A.J., eds., 1978, English for specific purposes: a case study approach. Longman, London.

Mackay, Ronald & Palmer, Joe Darwin, eds., 1981, Languages for specific purposes: program design and evaluation. Newbury House, Rowley, Mass.

Mackey, W.F., 1971, Free language alternation in early childhood education. Paper presented at the Conference on Child Language, Chicago, November 1971. (Eric Document Reproduction Service No. ED-060-755) (Cited in Nord 1981).

Matthews, Peter, H., 1979, Generative grammar and Linguistic Competence. Allen & Unwin, Winchester, Mass.

Munby, John, 1978, Communicative syllabus design: a sociolinguistic model for defining the context of purpose-specific language programmes. Cambridge University Press, Cambridge.

Nord, James, R., 1980, Developing listening fluency before speaking: an alternative paradigm. System 8: 1-22.

Nord, James, R., 1981, Three steps leading to listening fluency: a beginning. In Winitz, ed., 1981: 69-100.

Postovsky, Valerian A., 1974, Effects of delay in oral practice at the beginning of second language learning. Modern Language Journal 58: 229-239. Reprinted in Blair, ed., 1982: 67-76.

Postovsky, Valerian A., 1981, The priority of aural comprehension in the language acquisition process. In Winitz, ed., 1981 170-186.

Sapon, Stanley, M., 1971, On defining a response: a crucial problem in the analysis of verbal behavior. In Paul Pimsleur & Terence Quinn, eds., The psychology of second language learning, Cambridge University Press, Cambridge, England, pp. 75-85.

Scherer, George, A.C., 1950, The psychology of teaching reading through listening. German Quarterly 23: 151-160.

Scherer, George, A.C., 1952, The importance of auditory comprehension. German Quarterly 25: 223-229.

Straight, H. Stephen, 1976, Comprehension versus production in linguistic theory. Foundations of Language 14: 525-540.

Straight, H. Stephen, 1978, Consciousness as anti-habit. In K.D. Irani, Louise Horowitz, & Gerald Myers, eds., Pathology and consciousness: theory of consciousness, theory of mind, linguistics, Haven Publishing, New York, pp. 1-4.

Straight, H. Stephen, 1980, Auditory versus articulatory processes and their development in children. In Grace H. Yeni-Komshian, James F. Kavanagh, & Charles A. Ferguson, eds., Child phonology, Volume 1: Production, Academic Press, New York, pp. 43-71.

Straight, H. Stephen, 1982, Structural commonalities between comprehension and production: products of monitoring and anticipation. In Francis Lowenthal, Fernand Vandamme & Jean Cordier, eds., Language and language acquisition, Plenum Press, New York, pp. 177-180.

Terrell, Tracy, D., 1982, The natural approach to language teaching: an update. Modern Language Journal 66: 121-132.

Wilkins, D.A., 1974, Second language learning and teaching. Edward Arnold, London.

Wilkins, D.A., 1978, Notional syllabuses. Oxford University Press, Oxford.

Winitz, Harris, ed., 1981, The comprehension approach to foreign language instruction. Newbury House, Rowley, Mass.

Winitz, Harris & Reeds, James A., 1973, Rapid acquisition of a foreign language (German) by the avoidance of speaking. I.R.A.L. 11: 295-317.

Winitz, Harris & Reeds, James A., 1975, Comprehension and problem solving as strategies for language training. Mouton, The Hague.

ON THE REPRESENTATION OF ALGORITHMIC CONCEPTS

Elmar Cohors-Fresenborg

1. INTRODUCTION

In the following paper we will report from the aspect of concept representation on fundamental experiments, which we have performed during the last few years. These experiments ask how thirteen-year-old pupils form concepts in the area of algorithms and in which way they use these concepts in problem solving. Our fundamental hypothesis is that the central problem in programming is to organize a sequence of actions which has to be executed by the computer. From this hypothesis we were led to develop didactical material and a suitable curriculum, by which pupils may readily acquire a fundamental insight in central concepts of automatization and computer programming (Cohors-Fresenborg, 1976, 1978; Cohors-Fresenborg,et al. 1982).

If one follows the hypothesis that programming is a kind of organization, one may ask how important is language - as a meta-language for the structuring of problems and as a formal programming language for the representation of an invented algorithm. Our approach to create algorithms below the level of language in the form of simple acting sequences, and to introduce these by suitable didactical materials to pupils, has to be regarded in connection with ideas of Lowenthal (1982). Lowenthal has introduced the concept of non-verbal communication devices as a requirement for the possibility of introducing logical reasoning to young pupils by a non-verbal concept-formation approach.

2. CONCEPT FORMATION AND LANGUAGE IN MATHEMATICS

The common way of forming a concept in mathematics uses language in an essential way. One example is the introduction of mathematical concepts by nominal definitions. Another example is the equivalent form of representation of mathematical concepts in set theoretical notation (by the axiom of comprehension). Even in this case, the classical definition of a set $\bigwedge_{P} \bigvee_{y} \; y = \{x \mid Px\}$ uses elements of language (in the form of a predicate P) for the comprehension of the set which is to be defined. The other approach to define a set by enumeration is from the mathematical point of view a special case, but from the didactical point of view it has to be distinguished from the language-orientated way of defining sets. In the context of experimental psychology (or math education), the relation of elements and with this, the forming of the concept of a set, can be accomplished through a series of actions. A well-known example for such an approach is the introduction of fundamental set theoretical concepts and mathematical operations by means of the logical blocks of Dienes. In this approach, pupils first have a sufficient experience of examples by working with a small set given in the enumerated form, before their ability and pleasure for abstraction leads to the concept. A similar approach is often chosen in geometry: one starts with the handling of concrete geometrical or pictoral objects and then goes on to the mathematical object and concept through several levels of abstraction. Corresponding to those hierarchies of concept formation, there exist hierarchies of understanding and applicating the mathematical knowledge. In this context, we mention as an example Bruner (1966) (enactive, iconic, symbolic). The levels of abstraction by using geometrical objects are the basis for the hierarchy of van Hiele (1976).

It seems obvious from the logical and set theoretical point of view in foundations of mathematics (and in the didactical theories which in the last analysis are based on it), that acting with concrete objects is only a first step in the aquisition of mathematical concepts. The elegant availiability of these concepts is only given if they are represented at the symbolic level. In addition to logic and set theory, there is a third way of approaching mathematics by the concept of action. Although well-known in the foundations of mathematics for some time, this approach (namely computable functions) had only little influence on the discussion in mathematics education and the mathematical curriculum in schools. We will not consider here the consistent use of the operative principle in the primary school, e.g. the use of Cuisenaire rods. The impressive development of pure mathematics based on a bourbacistic view has obviously marked the way even for the didactical discussion.

We consider that it is useful to reconsider the role of actions in connection with our investigations of how pupils form fundamental concepts in the area of algorithms and in which way they use these concepts in problem solving. At a first sight, a computer program written in a formal programming language seems not very different from mathematical expressions used in set theory or pure algebra. But the use of computer program shows that the aspect of action plays a more fundamental role in this mathematical representation. To conclude, a computer program should organize the sequence of actions of the computer and not express truths or their logical connections.

Therefore, we have chosen (as a pupil-orientated introduction into fundamental concepts of programming) an approach through the organization of sequences of actions. The concept formation is more done by experience of acting than by a (nominal) explanation or definition.

One could expect that a sophisticated investigation of the pupils' behavior in using such algorithmic concepts would reveal other behavior patterns than those discussed until this point in the analysis of the process of mathematical concept formation. In the following, we first will show what we mean by the introduction of algorithmic concepts which is based on different levels of representation of actions. Then, we will describe which different cognitive strategies we have found during our investigations of pupils. In the following analysis, we will especially consider how we have to estimate the role of verbally-represented concepts in our experiments. The fact that we have proved our nearly non-verbal approach in constructing algorithms even with deaf pupils (Cohors-Fresenborg and Strüber, 1982) contributes an interesting aspect to this discussion.

3. DIFFERENT LEVELS OF REPRESENTING AN ALGORITHM

In the following we will show how algorithms may be represented in different ways. We use natural numbers as data-structure, the elementary operations of the processor are counting forwards and counting backwards. The control structures consist of the sequence of partial algorithms and the iteration (while a storage remains non-empty) of algorithms.

The Handling with Sticks

The elementary operations of counting forwards and backwards with natural numbers can be realized in a very evident way if one represents the natural numbers by a heap of sticks (match sticks) which one puts into the columns (on a sheed of paper). These columns represent the storage registers. Then, the elementary operations consist of the adding or taking away of a single stick in the respective column. In the following paragraph, we will develop in this level of representation (with match sticks) an algorithm for the addition of two natural numbers.

Let x_1 and x_2 be two natural numbers given in the storages (regisers) R_1 and R_2. The following algorithm computes the addition:

Repeat while there is at least one stick in R_2:
  Take away a stick from Register 2;
  Add a stick in register 1.

A protocol of the execution of this algorithm for the example of $x_1=3$ and $x_2=2$ looks as follows:

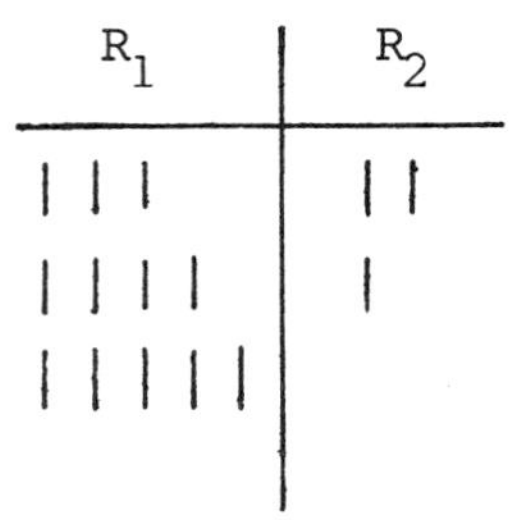

Figure 1.

In this protocol there is an extra line for every execution of the loop. We now give a second example, an algorithm for the subtraction process (if $x_1 \geqslant x_2$):

Repeat while there is at least one stick in R_2:
  Take away one stick from R_2;
  Take away one stick from R_1.

The representation of the algorithm on this level clarifies several aspects of the concept of an algorithm: The adding or taking away of single sticks underlines the aspect, that an algorithm describes a sequence of instructions which have to be executed step by step. This level supports the hypothesis that in order to construct an algorithm, it is important to organize a sequence of actions. The control of the algorithm by the stop decision (is R_i=0?) is represented in a very natural way.

Our classroom experiences and especially our case studies with pupils of age thirteen have shown that there are many pupils who prefer to develop an algorithm in this level of representation.

This idea was taken up by W. Back (1983) when he constructed his "Know-How" computer. By the presentation of this computer in

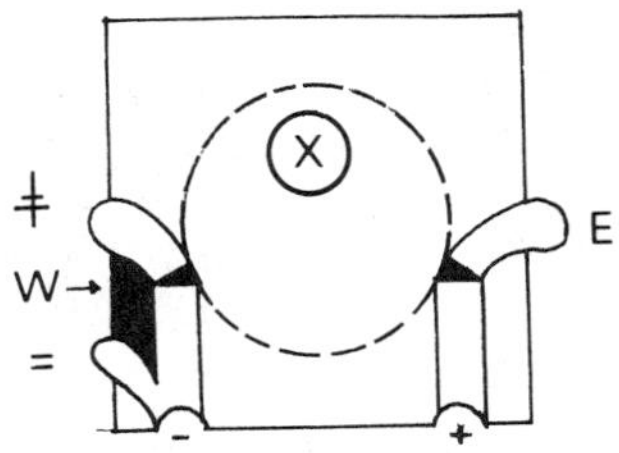

Figure 2.

the Computer Club of West German Television, it has become the most wide-spread "micro-computer" in Germany.

The Computing Networks

The presented level of representation by the sticks indeed simplifies the local invention of an algorithm, but has the disadvantage that after the execution, there is no documentation of the action, but only the result is shown. Therefore, we have developed (as the second level of representation) so-called computing networks, which we now present in detail. The columns which represent the counters on the paper, are replaced by the counter bricks (fig. 2) from the didactical material "dynamic mazes".

Within the counting bricks, there is a toothed wheel on which the numbers from zero to ten are written. Exactly one number is visible in the window x. The counter brick has two entrances, "+" and "-". If one drives with a pin through the "+" entrance, one reaches the exit E and the toothed wheel counts forwards. If one drives into the "-" entrance, two cases have to be distinguished: if the shown number x is >0, then the toothed wheel counts backwards and one reaches the exit ≠ ; if the visible number is =0 then the way to the toothed wheel is blocked by a wing, so that one reaches the exit = immediately without counting backwards. If one drives through the entrance "+" in the state 0, the wing W will be turned into the normal position as it is shown in fig. 2. In summary, one can say that with the counting brick, the storage of natural numbers can be realized. Furthermore, by suitable driving through the entrances the operations of counting forwards, backwards and the zero condition can also be realized. The realization of an algorithm in the level of the computing nets consists of connecting the entrances and exits of the counting bricks by rails (straights, curves, crossings, junctions) to a kind of railway network, so that the following happens: If one puts natural numbers into the counting bricks of the railway network and starts driving through the railway network at the entrance, one will reach automatically the exit and the result of the execution of the algorithm will

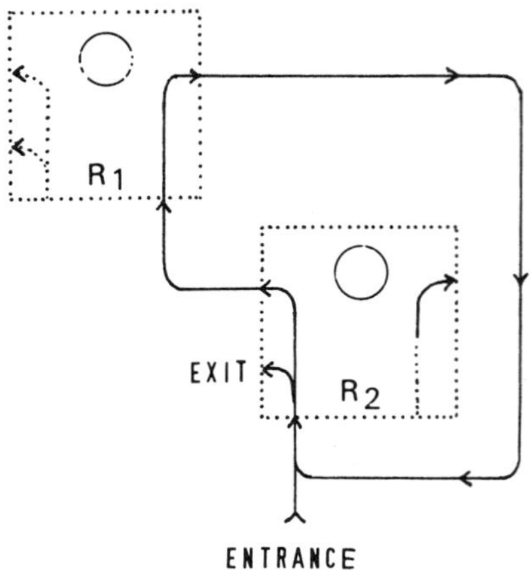

Figure 3.

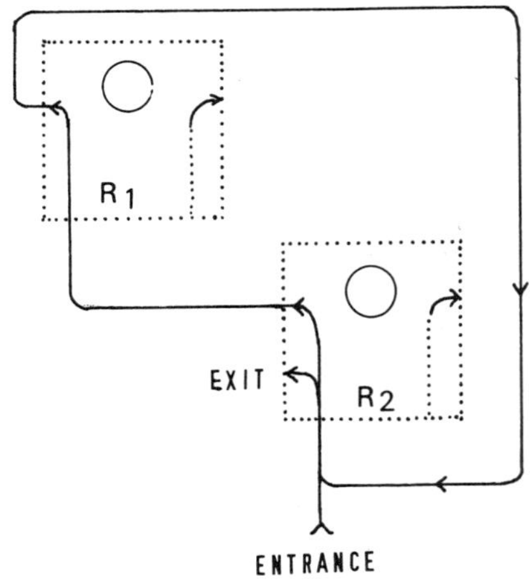

Figure 4.

appear in the counting brick which has been declared before. Figure 3 shows a computing network for addition, which in R_1 gives the result of the two numbers which were originally in R_1 and R_2. Fig. 4 shows a substraction network which computes the difference of the two numbers and displays the result in R_1.

The figures 3 an 4 show that there is a close relation between the graphic representation of such computing networds and flow charts. Therefore we present in fig. 5 and 6 the respective flow charts.

In these, A_1 denominates the counting forwards in R_1, and S_1 or S_2 the counting backwards in R_1 or R_2. The computing networks may be regarded as functionable flow charts.

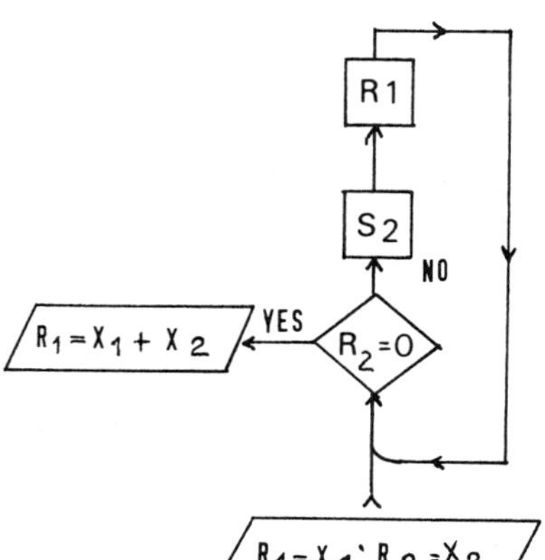

Figure 5.

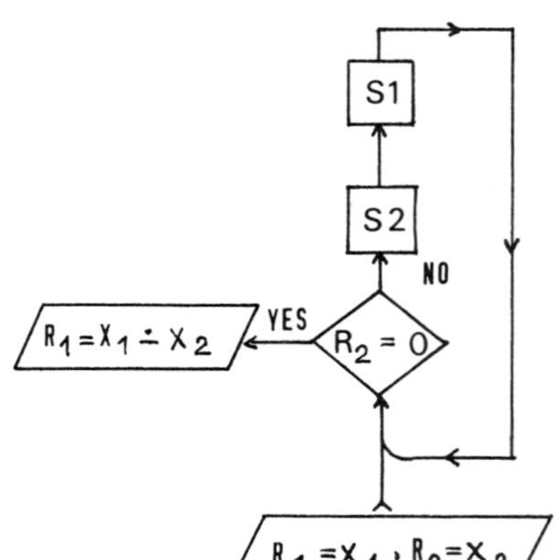

Figure 6.

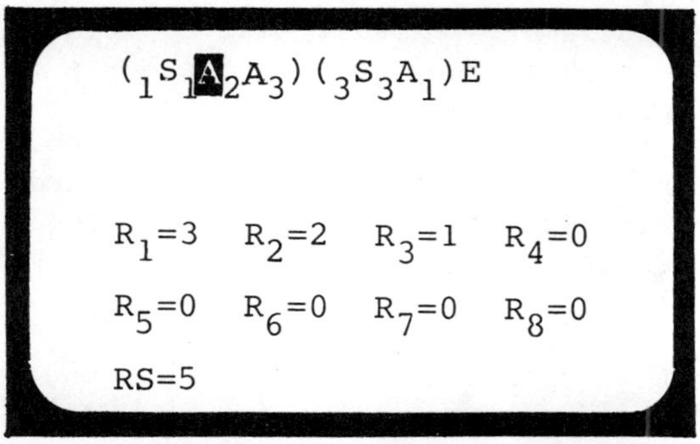

Figure 7.

Program Words for the Registermachine

To represent the algorithms in the symbolic level of a progamming language we have, on the basis of a micro-processor, developed a model-computer "Registermachine" [1] which has a very simple programming language and enables one to follow the execution of the programme and the change of the content of the registers step by step on the screen (fig. 7).

The language of the Registermachine is defined as follows:

Syntax:
1. A_i, S_i are program words ($1 \leqslant i \leqslant 9$)
2. If P and Q are program words, then PQ is also a program word.
3. If P is a program word, then $({}_iP)$ is also a program word ($1 \leqslant i \leqslant 9$).

Semantics:
1. A_i directs the computer to count forwards by 1 in Register i, and S_i directs the computer to count backwards by 1 in Register i ($1 \leqslant i \leqslant 9$).
2. PQ has the meaning: First execute P and then Q, "Sequence".
3. $({}_iP)$ has the meaning: While Register i is non-empty (that is, $R_i \neq 0$) repeat P ($1 \leqslant i \leqslant 9$), "Iteration".

Therefore, the language for the Registermachine is defined recursively by the elementary operations "counting forwards" and "counting backwards" with the help of the control structures sequentialization and iteration.

If one introduces Registermachines into the mathematics

---

[1]. In the meantime a Registermachine has been realized by a software system with a diskette for normal micro-computers.

curriculum of grade seven, where it will coincide with the introduction of the concept of functions on the basis of algorithms, one may define in a second step an extension of the language by introducing a suitable notation for subprogrammes. This notation is chosen in such a way that even recursive procedures are definable with it and that the execution of recursive procedures can be followed on the screen of the Registermachine as well (Cohors-Fresenborg, et al. , 1982).

The observation of pupils has shown that the representation of an algorithm as a programme word for the Registermachine creates no difficulties if it has been previously invented on the level of sticks or computing networks. Then it is only a question of translation from one "language" (the bricks, "counter" and the railway pieces are the words) into another [the words are A_i, S_i, $(_i,),$] . The handling of the Registermachine is also quite simple, and can be learned in a few minutes.

The algorithms for addition and subtraction which have been developed in the chapters "handling with sticks" and "Computing networks", are represented on the level of program words as follows:

$(_2S_2A_1)$ $(_2S_2S_1)$

On this symbolic level of the programming language, the analogy in both algorithms is obvious: The control by a loop is common, in one case the procedure of counting forwards is repeated, in the other case, the procedure of counting backwards is repeated.

As one observes the execution of the program word on the contents of the register which are indicated on the lower part of the screen (fig. 7) of the Registermachine, one realizes that this is an embedding of the handling with sticks into the level of the Registermachine.

The Idea of Variables

We have shown, how an algorithm can be represented on three different levels. Our approach contains especially an introduction of the idea of variables. The paradigma of a variable is:

- on the level of handling with sticks
  the uncounted heap of sticks, thrown into one compartment;

- on the level of computing networks
  that number which is in principle visible in the window, but covered;

- on the level of program nets
  the register, on which the program is working in one step (the actual values of the several variables are indicated in the lower part of the screen (fig. 7).

On the first two levels there is a non-verbal representation of the concept "variable". The idea of registers has two aspects: a non-verbal - if the working of the program is considered; a symbolic (and this is a kind of verbal) if the actual value is considered.

4. LEVELS OF REPRESENTATION AND THE PROCESS OF PROBLEM SOLVING

Our considerations about the three different levels of representation of algorithms (handling with sticks, computing networks, program words) lead to the conclusion, that the sequence of their presentation describes a hierarchy. One may have in mind even the hierarchy of concept formation which Bruner has described by "enactive", "iconic", "symbolic".

The idea of a hierarchy of levels of representation in a network of concepts indicates that pupils form these concepts most easily along this hierarchy. This implies the conclusion that there is an easier access in the lower level.

Our observations in the classroom and in many case studies lead us to the assumption that in the domain of our problems the three levels of representation which we have presented do not form a hierarchy in the above-mentioned sense.

In the following section, we will report some advantages and disadvantages which are connected by the choice of the different levels of representation with the process of problem-solving in tasks of the type "construct an algorithm".

The Handling with Sticks

Pupils which begin to invent an algorithm on the level of the sticks will normally use concrete numbers. If we consider the addition problem as a first example for algorithms then there exists the difficulty that the pupils don't understand the full meaning of the words "repeat, while..." as a control structure for an algorithm, because they are misleaded by the example with concrete numbers to repeat their actions until the desired result is reached.

The pupils are not aware of the zero-question, because it is obvious if a register does not contain any sticks. At best, they were aware of the zero-question in that moment when there were no sticks in the register.

Altogether, it is not easy to establish the idea in weak pupils that an algorithm has to be invented independent from the concrete example. But these problems can be avoided to a great

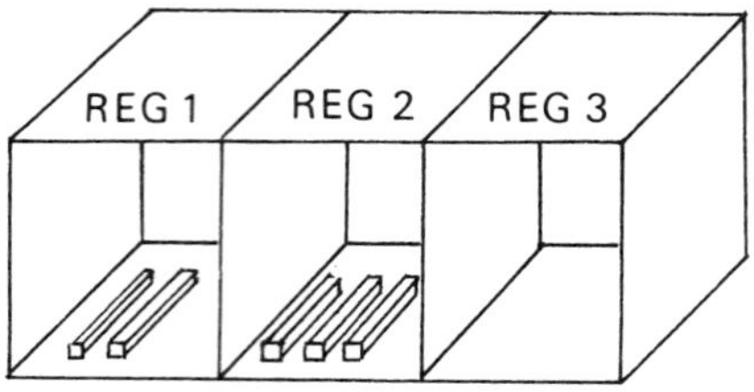

Figure 8.

extent, if the following change of design is introduced: Take a box with several compartments (fig. 8) into which the pupil can grasp but where he can't see how many sticks are lying in the different compartments. The idea of variables is stressed if the teacher grasps a heap of sticks with his hand and throws them into one compartment of the box. This heap of sticks corresponds to the mathematical phrase "Let x be an arbitrary but concrete number". The pupils don't have to abstract the general solution from several examples, but they can generalize it even from one example.

Even the zero-test is discovered evidently by the handling in the register-box: Before each stick can be removed, the pupil has to feel precisely whether or not there is a stick in that compartment.

By this improved design, there is still the advantage of the concrete operation and, on the other hand, there is, from the very beginning, a separation from the concrete number example.

The handlings with sticks seems to lay the intention during the phase of construction on the decision whether in the next step a stick should be added or taken away, and in which storage this operation should be performed. There are pupils who, especially in the beginning of our experiment, preferred this local reasoning and developed from this their problem-solving strategies of the construction of the algorithm. The handling with sticks has - in addition to the mentioned problem, that at the end only the result is seen and not the algorithm - the disadvantage that it is impossible to correct mistakes which are recognized during the elaboration, in a way that using a different strategy, one may get a correct result. Furthermore, the handling with sticks is easily disturbed by the discovery of new ideas or changes of the strategy. Normally, those interruptions or changes of the action sequence require the entire procedure to be repeated.

The handling with sticks must not necessarily guarantee that the pupil is aware of the structure of the developed solution of the problem. For example, it is not necessary that the pupil is aware that he has constructed an algorithm which consists of two loops which have to be executed one after another. On the other hand, one can see that pupils are grouping the actions, that means that they make a pause corresponding to the structure of the algorithm. By this, a kind of "meaning" into the sequence of doing is introduced. Thus, the doing is changed into a goal-orientated acting.

On the other hand, the handling with sticks can simplify a structuring on the level of the given problem. During the structuring process of the intended solution, the pupil may imagine for example that the heap of sticks has to be moved from one compartment into another one, or that as an interim solution there should be twice as many sticks in one compartment than in the other. Thus, the representation of numbers in the registers by a heap of sticks can simplify a method of problem solving, which involves the pupil imagining which partial solutions have to be reached. The development of an algorithm on the level of the sticks can certainly be connected with a conceptual analysis of the problem.

During our case studies, we have sometimes observed that pupils have developed (after several trials) a correct algorithm on the level of the sticks, but that they nevertheless did not know (when asked to repeat the solution or to write it down as a program word) how they really got the result. Still, this only happened when omitting the level of the computing networks. On the other hand, there were hardly any problems in translating a computing network into a program word for the Registermachine.

The representation of an algorithm on the level of the computing networks has a definite advantage compared with the handling with the sticks: At the end of the problem-solving phase, there exists on the board not only the result, but also a representation of the algorithm. Further, the necessity to connect the counters one after the other by railway lines stresses the idea that an algorithm consists of the organization of a suitable sequence of actions. We should especially remark that the iteration of parts of the program on the level of the computing networks has a remarkable representation: the repeated driving through the built-in loop is described by the pupils with the words "I am driving .... laps". Our observations of pupils have shown that the following interim form during the phase of inventing an algorithm is also suitable: On the board only the counters are fixed, the pupils drive (with their fingers) along the railway lines which only exist in their minds, or they draw the connecting lines by a pencil on paper, but they in fact change the content of the counters while driving through. This technique reconfirms the advantage of the computing networks compared with the flow charts: the change in the counters is "hands-on" experience. The advantage compared with a complete construction of the computing networks is that the time between the development of spontaneous idea and the completion of the built network is relatively short. Also, one is easier engaged in the provisional idea if its realization is not connnected with such constructive expense.

5. THE IMPORTANCE OF THE LEVELS OF REPRESENTATION IN THE ANALYSIS OF ALGORITHM

Up to this point, we were only concerned with problems in which one had to construct an algorithm. During our investigations, we also have given problems to the pupils in which they had to analyse the given algorithm. Normally, we gave the pupils a program word for the Registermachine to analyse. Only very seldom were they asked to analyse a computing network. The representation as a program word obviously makes the structuring of the program easier. One reason is that the representation in the symbolic level needs less space so that the algorithm can be grasped at once. The specific structure of the language for Registermachines supports reflections to disect a given program word into pieces. This can be done on the syntactic level with minimal use of semantics. Thus, a pupil can recognize such programs for which he already understands their method of operation. On the other hand, there exists a second possibility to grasp the meaning of the given program word: The pupil executes as a Registermachine the program word step by step. There are pupils who even use the sticks for this. After having some examples, pupils then formulate the problem in its general form (a term of a function).

The programming language for the Registermachine supports (by its simple but structured construction) the analysis of programs, because these have to be formulated even on the level of an algorithm in a clear structure (e.g. we do not allow arbitrary jumps). On the other hand, the pairs of brackets enable an easier structuring on the level of perception. This may be illustrated by a comparison with the language BASIC: On one hand this language enables (by the use of arbitrary jumps) even relatively unstructured algorithms to be formally expressed. If an unstructured algorithm is formulated in the language BASIC and it is difficult to analyse, this is due to the language only insofar as this language allows one to express chaotic algorithms. On the other hand, even the way an algorithm is formulated in BASIC does not support the structuring as the following example shows:
Fig. 9 shows the program in BASIC (only using counting forwards and backwards), fig. 10 the program word for the Registermachine.

On one hand, early all observed pupils prefered (in construction problems) a beginning on the level of sticks or computing networks, compared with immediately beginning with the program words. By this

```
 10  INPUT X
 20  IF X>0 GO TO 60
 30  IF Z>0 GO TO 90
 40  PRINT "X="X, "Y="Y
 50  END
 60  X=X-1
 70  Z=Z+1
 80  GO TO 20
 90  Z=Z-1
100  X=X+1
110  Y=Y+1
120  GO TO 30
```

Figure 9.

$({}_X S_X A_Z) \quad ({}_Z S_Z A_X A_Y)$

Figure 10.

they indicate that they regard this beginning as simpler. On the other hand, when confronted with analyzation problems, it is obvious that pupils had the fewest difficulties when the algorithm was given as a program word. Even a computing network for the problem of copying: $(x_1,0,0) \rightarrow (x_1, x_1, 0)$ is quite obscure, as the following comparison of figures 10 and 11 shows.

During constructing on the level of the computing networks, the long paths between the counters seem to be useful for those pupils who prefer a sequential strategy (Cohors-Fresenborg and Kaune, 1984). On the other hand, the long paths (especially when they are crossing each other), make it more difficult to obtain an

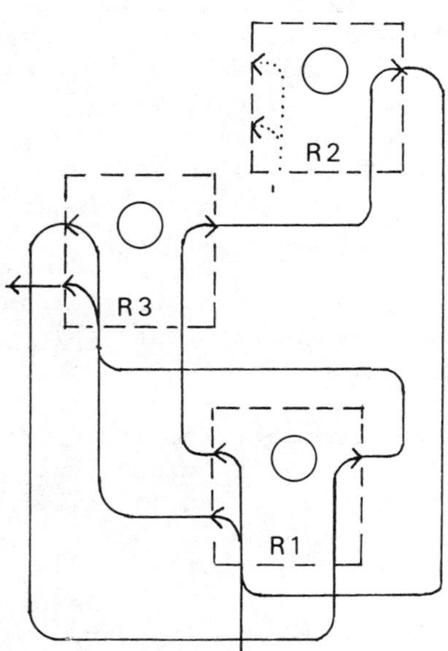

Figure 11.

$({}_1 S_1 A_3 A_2) \; ({}_3 S_3 A_1)$

Figure 12.

overview of the structure of the computing network during the analysis.

If one likes to present an analytic problem on the level of handling with the sticks, there are two possibilities: First the experimentalist could play an eaxample with the sticks. Secondly, one could follow the changes of the contents of the registers on the lower part of the screen, if the other part (where the program is to be seen) is covered. But it is hardly imaginable, that the pupil would prefer to analyse a given algorithm in this way.

6. PREFERENCES FOR DIFFERENT LEVELS OF REPRESENTATION

During the last few years, we have done quite a lot of case studies in which the pupils could choose on which level of representation they would like to begin constructing an algorithm. There is one group of pupils who prefers (for a longer time) a problem-solving approach in the sequence: sticks, counting networks, program words. In this approach, the difficulties occur when the pupil needs to invent a suitable sequence of actions with sticks. The translation of these actions into a computing network normally creates no difficulties. The translation of a computing network into a program is, in most cases, no problem.

There is a second group of pupils who begin on the level of sticks as well, but then omit the level of computing networks and try immediately to write down the program word. In this approach, there are more difficulties when they have to write down a program word for an action which they have developed and executed before. It is remarkable that in this approach, pupils often have difficulties expressing correctly the concatenation of two loops (e.g. the copy program fig. 12). But until now, the detailed analysis of this approach (and its observed difficulties) has not been completed.

We have the hypothesis that the level of the computing networks plays an important role in becoming aware of the structure which lies behind the invented sequence of actions.

Quite a lot of pupils start to invent an algorithm immediately on the level of the computing networks. It becomes obvious that their imaginations and the words which they choose to describe their ideas are closely related to the sequence in time of the elementary computations which have to be executed by the machine.

If a pupil has invented an algorithm on the level of the computing networks, he normally has no difficulties to translate it into a program word. But, there are only a few pupils who begin the invention of an algorithm immediately on the level of the program words. (Still, one must keep in mind, that our pupils are only thirteen years of age). It seems that these pupils begin with a conceptual structuring on the level of the problems, and later try to realize parts of the problem with already known subprograms.

Some pupils seem to have a sequential strategy (Cohors-Fresenborg and Kaune, 1984). We have the theoretical interpretation that they use a VMS-strategy (visual moderated sequence) during the development of an algorithm, in the sense of Davis and McKnight (1979). If the pupils begin the problem-solving process on the level of computing networks, this cognitive strategy is supported. It is an advantage for a VMS-strategy, that (by this approach) at the end of the problem-solving process, an algorithm is visible and not only the result of its application. But we have not yet finished our investigations on the problem of different cognitive strategies and of the existence of cognitive styles in constructing and analysing of algorithms. Our observations thus far indicate that the preference for one level of representation (which is shown in the beginning) is stable for an extended period. If one asks pupils to comment on the execution of the program by the Registermachine, one notices that some of the pupils use words derived from the description of the handling with sticks, while others use words which only make sense in the description of "driving" through a computation network. This indicates that the first approach in constructing an algorithm creates a paradigm which remains fundamental for a long time.

In our investigations on cognitive styles, we have seen that quite a lot of pupils prefer an approach in which the visualization of the acting of the machine (or the driving through the network) is the basis of their thinking. Others base their problem-solving process on the structuring of the given problem in a conceptual manner. These two different points of view play an important role in the discussion about programming languages: The answer whether PASCAL or LOGO is the more convenient and useful programming language may depend on the question, whether for a person's problem solving process it is more important to structure a given problem beforehand (or to organize actions (in a dialogue with the computer).

The choice of the level of representation in which the algorithm has to be invented has to be considered not only under the aspect of cognitive psychology. There also exists an interaction with regard to the content and the medium in which the algorithm is realized. Of course, it attracts attention that some solutions are more frequent in some levels of representation than in others. We would like to clarify this with the example of doubling: $(x_1,0,0) \rightarrow (0,2.x_1,0)$. There are two solutions which have the following representation as a program word:

I. $(_1S_1A_2A_2)$ II. $(_1S_1A_2A_3)$ $(_3S_3A_2)$

It attracts attention that nearly no pupil who begins in the level of computing networks tries to realize the first idea, but that many pupils who begin with the sticks do indeed choose this approach. When working with the sticks, it is obvious to replace each stick which has been taken away, with two new sticks.

The observed difficulty to proceed immediately to the development of a program word after handling with sticks could probably be hindered by the following support in the methodology.

After the algorithm is developed as handling with sticks, one could ask the pupil to repeat the algorithm and simultaneously construct the program word by using cards with the characters of the Registermachine's programming language. He then would lay the characters S and A with suitable indices one after the other and would nest these in a pair of brackets with suitable index as soon as he recognizes that he is repeating a sequence.

7. PROSPECT

Our considerations refer to different areas of research: One may ask to what extent the reasoning on algorithms is a mental activity which is connected with other mathematical mental activities. But the investigation of understanding and using algorithmic concepts can also be regarded as a special case of investigating mathematical concept formation, and can be subsumed under the discussion of theories about the concept of understanding. One may connect our investigations and considerations with theories of understanding and the hierarchy of the mathematical concept formation (for example van Hiele, Skemp, Herskowicz). Recently, Hasemann (1984) took first steps in this direction.

REFERENCES

Back, W., 1983, Der Know-How Computer, <u>in</u>: Mikrocomputer 5/1983, Franzis-Verlag, München.

Bruner, J.S. et al., 1966, Studies in Cognitive Growth, p. 1-67, New York.

Cohors-Fresenborg, E., 1976, Dynamische Labyrinthe, <u>in</u>: Didaktik der Mathematik 1/1976, p. 1-21.

Cohors-Fresenborg, E., 1978, Learning problem solving by developing automata-networks, <u>in</u> Revue de phonétique appliquée, no. 46/47, p. 93-99.

Cohors-Fresenborg, E., Griep, M., Schwank, I, 1982, Registermachines and functions, A School-book introducing the concept of function on the basis of algorithms. Osnabrücker Schriften zur Mathematik Series U, book 22E, 22 LE.

Cohors-Fresenborg, E., Kaune, C., 1984, Sequential versus conceptual - two modes in algorithmic thinking. in : Proceedings of the 8th Conference for the Psychology of Mathematics Education, Sydney, p. 261-267.

Cohors-Fresenborg, E., Strüber, H.J., 1982, The learning of algorithmic concepts by action. A study with deaf children; in: F. Lowenthal, ed., Language and Language Aquisition. New York: Plenum Press, p. 95-106.

Davis, R.B., McKnight, C.C., 1979, Modeling the processes of mathematical thinking. In: J. Child. Math. Behav. vol. 2, no. 2, p. 91-115.

Hasemann, K., 1984, Analysen mathematischer Lernprozesse mit Kognitionstheoretischen Modellen; inaugural disseration (unpublished), Universität Osnabrück.

Hiele, P.M., van, 1976, Wie kann man in Mathematikunterricht den Denkstufen Rechnung tragen? In: Educational Studies Math., vol. 7, p. 157-169.

Kaune, C., 1984, Kognitive Strategien von Schülern beim Programmieren; In: Informatik als Herausforderung an Schule und Ausbildung W. Arlt u. K. Haefner (Hrsg.), Springer, p. 241-245.

Lowenthal, F., Marcq, J., 1982, How do children discover strategies? In: Proceedings of the 6th Conference for Psychology of Mathematics Education, Antwerpen, P. 156-161 and p. 287-292.

The box of building bricks "Dynamische Labyrinthe" (Dynamic Mazes) is produced by Beschützende Werkstatt, Industriestr. 7, D-4500 Osnabrück, W. Germany.

A STUDY OF PUPILS READING GEOMETRY

M. Guillerault and C. Laborde

Laboratoire de Didactique des Mathématiques
et de l'Informatique
Université de Grenoble 1 France

The study presented here is based on an approach to the teaching of mathematics which takes linguistic problems into account. In this approach, these linguistic problems are considered in relation to the acquisition of mathematical knowledge within the classroom situation. What is required, we suggest, is not merely an analysis of learner utterances, teacher utterances or the wording of the text book. These utterances can only be fully grasped if they are related to their conditions of production, to the knowledge structure of the speaker or listener and to his relationship with others during the exchange. In other words, the discourse situation, as defined by A. Culioli (1976) cannot be ignored. One of the functions of the discourse situation is the transmission of meaning, meaning which derives from the cognitive constructs of the subject and which cannot be totally defined without reference to the context of the utterance. The meaning is constructed by the locutor/speaker and reconstructed by the interlocutor/listener through the referential values in the context of the utterance... . In the previous symposium, "Language and Language Acquisition", the importance of allowing for the situational constraints of language exchange, and the essential role of speaker and listener in the creation of meanings within this situation were stressed by M. Brossard (1981), and, at the end of the meeting, an appeal was made by F. Lowenthal (1982) for a development of contextual linguistics. The approach to the problems raised here is related to that analysed by J.B. Grize at the present congress, namely the representation of a microworld (which Grize calls schematization) elaborated by the speaker and reconstructed by the listener in function of his prior knowledge. The readjustement and adaptation resulting from this reconstruction are driving forces in the growth of knowledge.

It was within such a framework that an experimental communication situation was devised which included constraints enabling us to analyse, on the one hand, the interrelationship between conceptions that learners construct of the mathematical objects employed in the task, and on the other, either the verbal translation of these objects (speaker activity) or the decoding of the formulations concerning these objects (listener activity).

This experiment was the subject of a paper read at the previous symposium (Guillerault, Laborde, 1980). In what follows we recapitulate the experimental design and the main results that we were able to draw. We then go on to present the reasons which have lead us to extend the study by a second type of experiment.

FIRST EXPERIMENT

1. Recapitulation of the Experimental Design

Groups of two learners (encoders) were presented with figure 1. The figure was printed on a white sheet of paper with neither title nor heading.

Each group was informed that it had approximately one hour in which to work out and draft a mutually acceptable message which contained no figure and which would allow learners of the same age (11-13) to reconstruct the given figure perfectly. The main phases in the elaboration of the message were noted by an observer and the discussion between the two encoders was recorded. Once the message had been completed, and if the encoders had not already done so of their own initiative, the observer asked them to check their message by drawing the figure, using the text that they had just written. This phase was the emission phase of the message and was completed by one hundred pairs from approximately 30 schools in the Grenoble region.

The decoding phase by pupil pairs was carried out by a lesser number of pairs. Each group of decoders worked in the same conditions

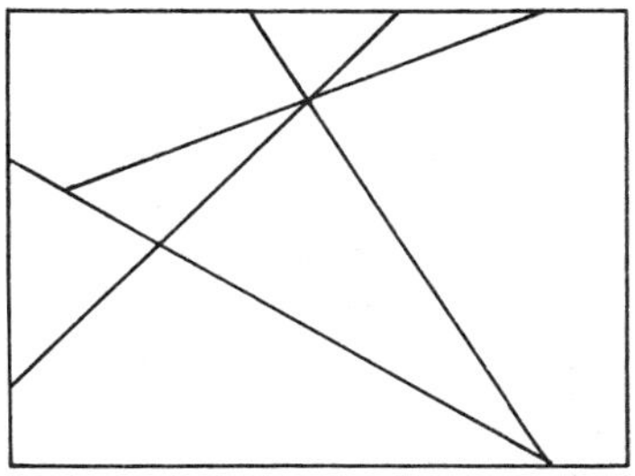

Figure 1.

as the encoders and without any communication with them. Their task was, with the sole aid of the message, to reconstruct the figure.

2. Recapitulation of Main Results

a) Encoding phase. The links between the conceptual analysis and the message form became apparent in different ways.
- Initially the description of the figure was sometimes no more than a static inventory of the objects of which it was composed, no indication being given of the interrelationship which might allow them to be located.
- Only three pairs did not go beyond this phase. The other pairs moved onto an instructional description (instruction for the measurement, the location, the drawing of the figure, the lettering or numbering of a point or segment). There was, at this stage, for many of the encoders, a further evolution in the description of the figure. First of all they attempted to describe the figure in terms of the juxtaposition of zones. Confronted by the geometrical and linguistic difficulty of such a project, they then tried to describe the figure by substituting for the zones the perimeter edges of the rectangle and the 4 interior segments. The result was that the segments were often detached from the zones that they bordered.

Likewise, a similar procedure was employed by certain pupils for the points which were the extremities of the segments. Then they were "detached" from the segments, initially expressed as material entities and described dynamically. For example: "2 cm away, towards the left another line starts and it finishes towards the right at 7.1 cm, downwards where you are measuring from".

It was not until later that the notions of "starting" and "finishing" of a segment or a line were replaced by the notion of point. Thus, from the evolution of the different encodings it was possible to detect a hierarchy; zone - segment - point. This hierarchy, it should be pointed out, is not unrelated to the presentation of the concepts of "point", "line" and "surface" in the old classical mathematics text books. Conversely, it is completely out of phase with the teaching of a "set approach" for these notions. It was, in fact this approach that the pupils in our experiment had been exposed to.
- This evolution, towards the recognition of a point or a segment as an object frequently occurs interdependently with the numbering or lettering process. (That is to say the labelling of the elements of the figure, either segments or points) (Laborde, 1982).

Of course, not every pupil pair went through this evolution in its geometrical analysis. For some, the evolution was partial, either because they went straight into the task at an intermediary (or even terminal) stage, or because, on the contrary they stopped before reaching the ultimate stage.

As a result, there was a great amount of implicit information in the majority of the message formulations (points or even segments that were not expressed) and the inappropriate use of words. For example,

- "edge", "base" or even, "triangle" and "angle"; words related to the notion of zone were used to designate segments; "corner", "tip" were used to designate points.
- "end", "starting point", "top", "bottom", words related to the notion of segment were used to designate points.

b) Decoding phase: The most striking aspect of this phase was apparently arbitrary decisions of the decoders when confronted with the message formulation. The decisions, in contrast with the encoding stage, were often taken without discussion between the partners as to the possibility of alternative choices. The decoding analysis also turned out to be more tricky than that of the encoding stage. This was because decisions taken at any given moment were motivated not only from the reading of a certain formulation but also, from all the interpretations of the preceding part of the message, and more particularly, on the state of reconstruction of the figure. (Guillerault, Laborde, 1980).

Once the experiment had been finished, it became clear to us that a certain number of the decoder difficulties such as the fact that the message was handwritten, the lack of punctuation, the lay out, the unduly high amount of implicit information, the muddled nature of the formulations, made the decoding analysis sometimes needlessly complex. Furthermore, working in pairs, which had been so profitable during the encoding phase of the message had provided less information and had given rise to less interaction between pupil pairs during the decoding stage. The experimental design was, compared to the information that it provided, too heavy. We thus decided to simplify the design and to use this simplification as a way of increasing the numbers of our sample.

DESCRIPTION OF THE SECOND EXPERIMENTATION

1. Choice of Message Samples

Of the original one hundred messages only about thirty were retained. These were messages which contained, at a maximum, one or two ambiguous statements and which were reasonably clear to understand. After essential punctuation had been added and the text had been, where necessary, divided into paragraphs, the message was typed out.

2. Decoding. Procedure

During a lesson each pupil was given a message to decode.

After drawing the figure he was given the original figure and was asked to correct what he considered to be the "errors" in the text of the message. This procedure of working alone in class (each learner working on a separate message) made it possible to collect eight or nine decodings for each of the selected messages - producing a total of 247 decodings. The experimental design (variously coloured felt tipped pens, replacement sheets of paper... were provided) enabled us to keep track of the rectifications and over-drawings as the work progressed.

THE ANALYSIS. AIMS AND LIMITS

In this experiment the decoding of the message conditions a second activity, namely, the construction of the figure. It is through the results of this second activity that the comprehension of the decoder is evaluated. The data gathered is used to help in the interpretation of the discrepancies between the original figure and the figures produced by the decoders. These differences are attributed to different decoding deriving from the following parameters:
- The conceptions that the learner has of the objects and of the geometrical relations underlying the figure.
- The decoder's conception of the task that he is being asked to perform, and, more particularly, the image that he has of the writers of the message that he has to decode. These are school children of his own age, who do not necessarily express themselves, "as well as teacher does". (Note: it was important that the messages offered for decoding should be of sufficient clarity to avoid a refusal motivated by incorrect wording of the message).
- The interdependencies which arise between the drawing task and the decoding task: what actually goes on during the decoding process depends very much on the aim that is being pursued: for example, when the aim is the repetition of the text, memory will play a primary role - a role not at all called upon in this experiment.
- The actual reading of the text: this includes the interpretation of the overall organization of the text, the phrase structure, word groups or lexis.

Quite clearly, these different parameters come into play in conjunction with each other and it would be pointless to try and explain a given reading behaviour by taking account of one alone. For simplicity's sake however, we have interpreted certain cases by affirming the primacy of one or other of these parameters. Examples of this will be given below.

It should be stressed, however, that the interpretations that we advance, and especially those concerning conceptions, are based essentially on the analysis of the encoding phase (1st experiment). This analysis enabled us to pin-point some of the learner's geometrical conceptions which underlie implicit or ambiguous information in the message. If these ambiguities or implicit messages "get through" during the decoding phase (in other words, if the decoding is carried out as was inteded by the encoders) we then assume that the decoders share the same conceptual framework and it is from this conceptual framework that the interpretation of the message derives.

Our evaluation of the decoding stage may appear somewhat indirect as it is based on a subsequent drawing activity. It is true that our analysis doesn't enable us to capture the comprehension of the message totally. We are not always in a position to know precisely what the decoder has understood by different lexical items or how he has understood the structure of a complex nominal phrase. And of course it is true to say that the drawing that he produces is not merely a function of the corresponding phrase in the text but it also depends on both the text as a whole, as well as factors outside the text. One of the most important of these factors is the state of the figure as it has been drawn up to then. This can either hinder a certain construction or, on the other hand, favour a different one. It must be born in mind that what the learners consider as "normal" geometrical figures is governed by implicit rules: For example, segments should not be too close to one another; if they cross each other the points of intersection should fall within a reasonable area.

However, we would maintain that all evaluative methods of a decoding activity have limitations and necessarily privilege certain aspects. Thus, the repetition of the text privileges memorisation of data more than it does the capacity of processing the data after decoding. (Rasolofoniana, 1983 pp. 24-25). To say this does not mean to imply that comprehension and memorisation are completely separate phenomena. Methods of observation of decoding stages, where learners are required to read and reflect aloud, modify the learners' natural behaviour (Walther, 1981).

The method we use enables the observer to monitor the results of the decoding of written information at the moment of transfer into a different mode of representation. In so far as our analysis takes this framework into account, and situates the decoding activity within a communicative context, with all its various parameters, it seems to us to supply significant data for the study of decoding behaviour of learners at this age.

SELECTED EXAMPLES

In what follows, the decoding of 3 messages, purposely selected for their very different natures, are presented [1] . The first two (messages A and B) contain certain features by virtue of which they diverge from standard mathematical discourse. (Further clarification is given below). The third message, message C, is, by contrast, both in terms of the lexis and of the symbolism used, very close to textbook discourse.

Message A.(24 decodings) (Figure 2.)

"The coordinates of the rectangle 8 cm long and 6 cm wide. Starting from the left side at the top, mark a point 3 cm away called A, B is 2 cm away from point A, C 2 cm from point B. There is 1 cm left.
On the left hand side of the long side at the bottom make a point at 7 cm called D. There's 1 cm left.
On the left hand side of width starting from the top, make a point 2 cm away from the edge called F, another point starting from F 3 cm away called E. 1 cm is left.
You join the straight line F and A to D. On the line F, which you've already drawn, take 8 mm and call it G. Join B to E, C to G and now check it. The line B and C cross each other on the line A at 1,5 cm. The line B crosses the line F and A".

(Sophie and Myriam)

Taken at the level of the geometrical objects described, the points in this message appear to be completely detached from the segments which they bound and thus have an autonomous existence. This is stated explicitly in the text (with exception of G) and is reinforced by the attribution of a code, in this case a letter, to each of them. However, it will be noticed that the base on which these points lie (that is to say the different sides of the rectangle) are not explicitly mentioned.

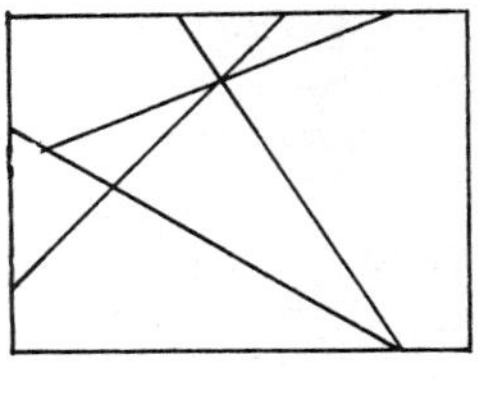

Figure 2. The figure is included here for the sake of clarity only.

[1]. An English translation of these message is given here; we have tried to translate the pupil's way of formulating.

There is, in this message, a further ambiguity. This derives from a particular conception of the geometrical object, segment. This ambiguity is shown by the fact that a straight line was labelled in the same way as the point which was initially chosen to mark its extremity. Hence, mention is made of "the line A", "the line F" and even "the line F and A"; the intended meaning of this last phrase being, "line F" and "line A".

As far as decoding is concerned, the same expression "on the left side" is twice repeated in the message. The first time it has the meaning of location on a base (on the lower length on the left hand side), the second time it is used to indicate the base (the lower length which is on the left). This raises the problem of the influence on decoding of multiple meanings of a given expression.

Hence, we can say that what fundamentally characterizes message A (with reference to the orientation of the rectangle on the sheets of paper and the base, on which the points are plotted) is two implicit facts:
- an ambiguity at the level of the designation of geometrical objects. (The same letter referring to either a point or a segment which has this point as its starting place).
- a polysemic expression.

Nevertheless, decoding for the message are good:

a) The implicit information about the rectangle's orientation was understood in 23 cases, with only one decoder persisting until the end in drawing on the paper a figure of a rectangle in an upright position.

b) The implicit as to the "base" of the points to be plotted (the sides of the ractangle) was understood in 21 cases. Only 3 decoders positioned the points A to G more or less arbitrarily respecting no more than the distance between the points. It is as if the figure described in the message has lost its uniqueness and has become no more than a contingent expression of a group of figures.

In our examination of point C) and d) we will restrict ourselves to the 20 decodings in which the initial implicit information was "correctly" understood. Amongst these, one pupil produced, with the exception of one axial symmetry, a completely correct reconstruction of the figure. (He systematically mistook right and left).

c) The questionable expression: "Join the straight line F and A to point D" produced the following decodings:

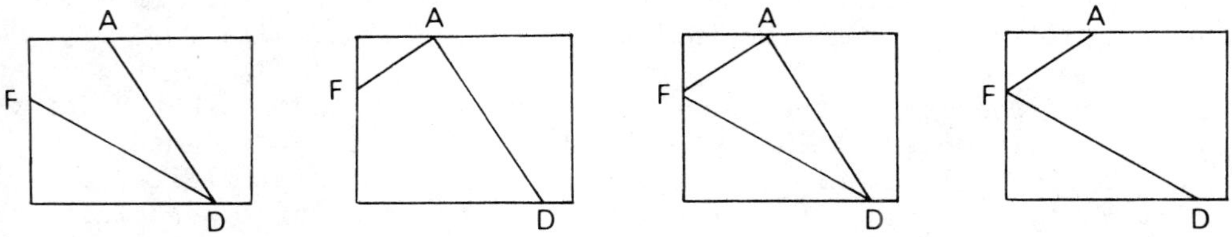

13 cases ("correct decoding) 5 cases 1 case 1 case

Figure 3.

d) As the figures above show, the polysemy of the expression, "on the left side", caused no trouble to any of these 20 decoders and it played a role in no more than 2 of the 4 other decodings.

Finally, it should be noted that out of the 5 decoders who constructed the 2nd of the above figures, one interpreted "Join B to E and C to G" as "Join B to G, E to G and C to G".

Message B (25 decodings) (figures 4)

"There is a geometical figure in a rectangle measuring 8 by 6 cm. At its base there are 3 points starting on the left. The first is at 1 cm from the edge, the 2 nd at 2 cm from the first and the 3rd 2 cm away from the second. From the 3rd point there's a line which goes up to a 4th point on the length, 1 cm away from the edge, starting from the left. From this point, No 4, there's a line which goes to a point 5 which is situated 2 cm away from the edge starting from the bottom on the right and going up on the wide side. On this line 5, 1cm left from the edge there's a line which goes to point I. At point 2, there's a line going to point 6 which is 1 cm from the edge on the width at the top on the right. You can check if the figure is correct by seeing if the 3 points of the base cross to make a point 7".

(Eric and Isabelle)

This message differs from the previous one in as much as it doesn't use letters to designate the points or segments. However

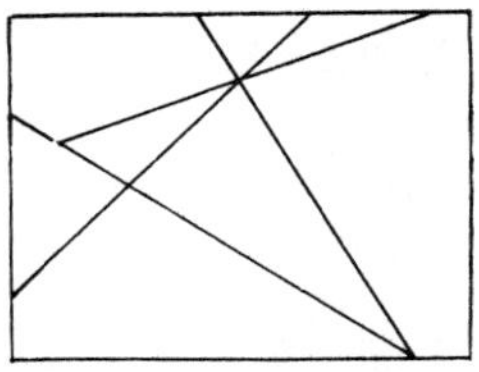

Figure 4.

there is an evolution of the system of location of points from the use of ordinals, at the beginning, to a numerical coding.

The location of the segment extremities produces cleft relative clauses or complex nominal groups. As in message A, a segment ("line five") is labelled in the same way as its extremities. But conversely to message A, where the segment extremities were initially plotted prior to the instruction to draw each segment, drawing the segments takes precedence over the location of their extremities and it is this, one might add, which gives the dynamic flavour to the message.

This then, is a message which seems, "a priori", to be difficult to decode. Nevertheless, out of the 25 decodings, 8 are correct and 6 almost so. That is to say they contain no more than 1 error (5 of these were in the location of a segment extremity and one in the interpretation of "at its base"). Thus the designation "line five" for the segment, one of whose extremities is point 5, was interpreted in 16 of the messages as the encoder meant it to be.

On the other hand, the following gave rise to interpretations other than those wished by the encoders:
- The expression "at its base" resulted in 5 decoders locating the points on horizontals other than the base length. In two cases these were arbitrary points.

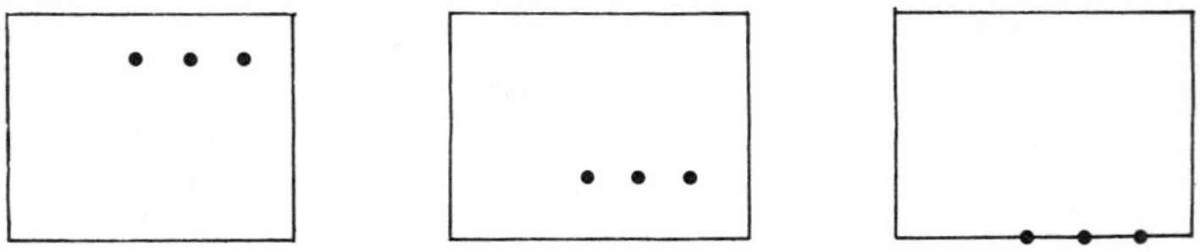

figure 5

- In the expression "on this line 5, 1 cm left from the edge", "left from" was divorced from its head "edge" and was interpreted as a way of locating on the segment.
Instead of the interpretation:

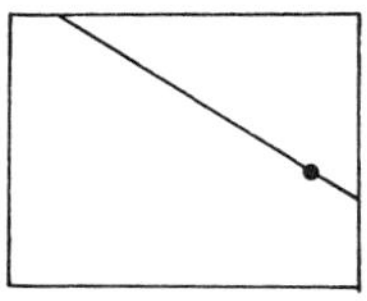

figure 6.

The following interpretation was made:

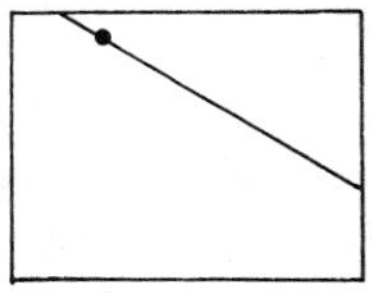

figure 7.

- Similarly, phrasing attributed to "Point 6 which is 1 cm from the edge on the width at the top on the right on the wide side" was given the following interpretation by fair decoders

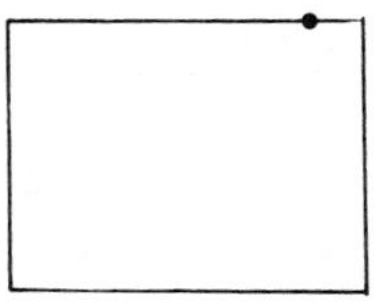

figure 8

(1 cm from the edge of the width) (at the top on the right) instead of

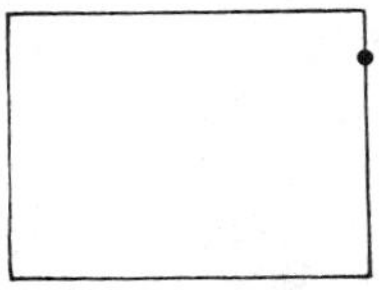

figure 9

(1 cm from the edge) (on the width at the top on the right).

- In one case even the punctuation of the text was completely altered. The phrase "1 cm away from the left edge there's a line which goes to point 1. At point 2, ...," is understood as meaning the drawing of a line starting from a point (point 5) going to point 1 and ending up at point 2.

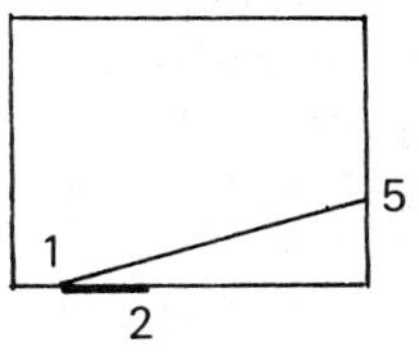

figure 10

Message C (9 decodings) (4 possible positions)

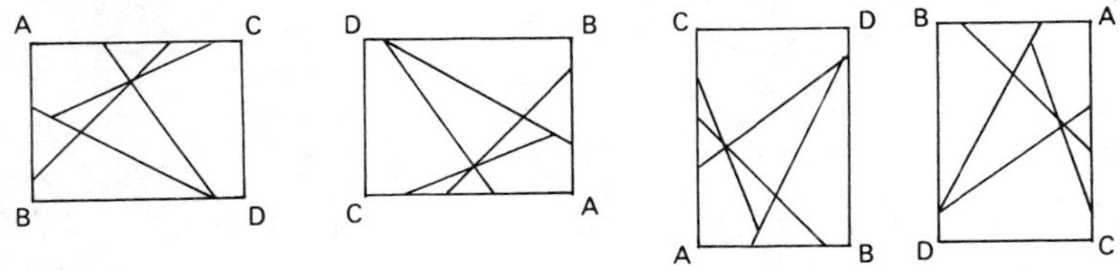

figure 11

"Draw a segment [AB] 6 cm long and then another segment [AC] so that [AC] is 8 cm long and lies perpendicular to [AB].
Draw a segment [CD] measuring 6 cm so that [CD] lies perpendicular to [AC] and then draw a segment so that B and D are its extremities.
Plot a point E, on the segment [DB], 1 cm away from D and then a point F, on the segment [AC], 3 cm away from C.
Draw a segment whose extremities are H and G and then plot a point I on the segment [AC] so that [AI] is 3 cm long. Draw a segment whose extremities are I and E.
Plot a point K on the segment [FE] so that [FK] measures 1 cm.
Plot a point J on the segment [CA] so that [CJ] is 1 cm. Draw a segment whose extremities are K and J".

(Nadine and Sheila)

Here the vocabulary and the use of symbols are the same as those used in maths textbooks. The message thereby differentiates itself, not only from the two preceding messages, but from all the other messages that we collected. In particular this message is the only one to use brackets [ ] for the segments.

After sometimes several false starts 7 decodings produced a figure which was either identical with the original or symmetrical to it. (The reason for this latter alternative being that there was no indication as to whether the vertices A,B,C,D should be distributed in a clockwise or an anti-clockwise order).

More surprising, however, are the two other decodings:
The construction of AC was understood relatively independently of AB. That is to say, the fact that the two segments had a common point A, indicated only at the level of the codings [AB] and [AC], was neglected. It seems likely that the presence of the adjective "another" in the expression "and then another segment AC" was interpreted as being of more significance that the fact that the same letter A was present in the two codings of the segments. (It is however very common in the teaching of mathematics, in so far as the objects appear in different exercises, to label two different objects with the same letter. Here the adjective "another" may have functioned as a boundary marker).
Here are the two decodings:

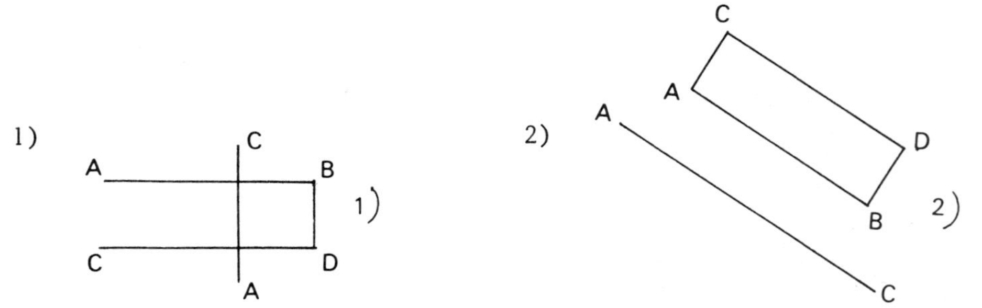

Figure 12.

However, in this case, the independence of the constructions is only partial. This can be seen by the fact that one of the two decoders respected the onthogonal relationship of [AC] and of [AB] while the other imposed a relation of parallelism between [AC] and [AB].

However, what is perhaps more surprising than the interpretation itself is the presence of two distinct points A and two distinct points C in the same figure. Despite this neither of the two pupils appeared to be in the least ill at ease with the ambiguity that had thus been introduced into the figure. In decoding 2) the segment AC, drawn outside the rectangle ABCD, was never used again. In decoding 1) only one segment AC is realised in the figure and it is this segment alone which, (with the exception of the construction of the point H) was used in what followed.

We have spent some time in the discussion of these two cases because they demonstrate the difficulties that learners can encounter when decoding texts written according to accepted textbook conventions - texts which, in the eyes of mathematicians, contain neither ambiguous nor implicit information. The difficulties that pupils encounter stem from current language usage in mathematics, usage which, precisely, is never explicitly formulated. An important example of this is the fact that two identical labellings can denote two different objects but, when this is the case the range of the first labelling stops where the second begins. On the contrary, mathematics teaching stresses the fact that two denotations can refer to the same object.

REVIEW OF PROBLEMS ENCOUNTERED

Quite clearly, it is not feasible within the limits of this article to make a comprehensive review of all the problems encountered either during the decoding of the thirty different messages or even during the decoding of the messages A, B and C, which we have already discussed in length in the preceding paragraph. Nevertheless we will now make some attempt to summarize the most striking decoding problems.

Decoding of Geometrical Objects

Conceptions concerning the geometrical object segment differ. We have pointed out elsewhere the confusions in the pupil's mind between the notions of point and segment (Laborde, 1982). Here, in the messages A and B, this conceptual confusion is demonstrated by the fact that a straight segment is designated by the same term as was initially chosen for plotting its extremity. It can be seen (in the two examples discussed above, as in others) that this ambiguity demonstrated here by labelling ambiguity is, in most cases, understood by the decoders in the same way as it was intended by the encoders. It would seem, therefore that this geometrical sort of ambiguity which one could summarily categorize as a "point/segment confusion" does not appear to the student so inacceptable as it would do to an adult (mathematician). Pupils are prepared to accept without qualms that a single symbol should designate two objects which are not only distinct (as the two points A and C in the decodings of the message C, analysed above) but which even belong to different categories of mathematical objects such as point and segment. This tendency is demonstrated not only by the acceptance and "correct" decoding of messages containing obvious ambiguities (messages A and B above) but in the creation of an ambiguity in the drawing of a figure from a non-ambiguous message, as message C above.

At another level, it is worth mentioning the difficulties arising from purely geometrical terms. However these terms seem relatively well understood by the decoders even in case of expressions which are ill-suited or infrequent with the single exception of the word "edge" ("bord" in French) whose polysemy was a frequent cause of misinterpretation of the author's intentions.

The Evolution of the Perception of the Task

Each decoder is aware of the fact that the message that he has in front of him is the description of a defined figure which he must reconstruct himself. However, quite commonly a drift was detected in the decoder's comprehension of the task. Certain data are taken into account but other elements, which may have been formulated in the message with a greater or lesser degree of clarity, are overlooked. This means that certain geometrical elements are located arbitrarily by the decoder who has thus "forgotten" the uniqueness of the figure described in the message.

A further aspect of this decoder comprehension drift is demonstrated by the fact that in certain cases priority is given to his own development of his own drawing of the figure. This can lead to decisions being made that are in flagrant contradiction with the text of the message. In this way "awkward" bits of the message can be, more or less deliberately, omitted.

Finally, although this is not the case for the messages that have been discussed here, we found some decoders who, once their (incorrect) figure had been completed, and when confronted with the original figure, proceeded to correct the proposed message so as to make it fit their own drawing.

Comprehension of Polysemic Terms

In this experiment we tried to determine how the different

meanings of a single term might influence decoding. It became clear that the "elastic" use of the same decoding term or expression to indicate different objects did not automatically entail particular difficulties during the decoding phase. We have already discussed the use of the same symbol referring alternatively to a point and a segment, and we have referred to the occurrence in message A of the same expression "on the left side" with different meanings.

Of course, the decoding of expressions, whose meaning varies at different moments of the message, depends, above all, on the context, on the state of the figure already constructed by the decoder and the potential phrasing that can be given to the text during the decoding phase. But, quite apart from the accuracy of the interpretation, it can be seen that the existence of such expressions with double or triple meanings is not, "a priori", a hindrance to decoders.

However, conversely it is also common to find expressions whose meaning remains constant throughout the message but to which the decoders attribute successively different meanings.

Decoding of the Text

During the decoding of the message the different phrasings of the text sometimes produced interpretations quite different from those intended by the encoders. Examples of this were mentioned during the analysis of the decoding of messages A and B: unjustified linking of terms separated by punctuation in different phrases, idiosyncratic decoding of the expressions of the message.

This sort of example is extremely frequent. We were also able to collect a considerable number of cases where a term such as "on the right" applied to a side of the rectangle is not interpreted as such, but its influence permeates to an ulterior part of the phrase and it is used to locate, on the right, some completely different elements to the one referred to by the encoders.

There have even been cases where a term appearing only once in the message has been used twice by the decoder, and each time with a different meaning.

CONCLUSION

We have not attempted to supply an exhaustive inventory of the geometrical and linguistic problems created by the decoding of the selected messages. In particular, we have not touched on the decoding of the French prepositions "sur" "de" "à" which are strikingly polysemic.

Even if this study has only been concerned with formulations produced by pupils, it nevertheless seems to us that we have demonstrated certain types of behaviour that are liable to be encountered when pupils are confronted with text-book or teacher discourse. What is required is further experimentation along these lines in order to be able to pin-point with greater accuracy the difficulties of the comprehension of mathematical formulations which teachers so frequently encounter.

REFERENCES

Brossard, M.,1981, Situation et signification: approche des situations scolaires d'interlocution, Revue de Phonétique Appliquée, n° 57.

Culioli, A., 1976, Transcription du séminaire de DEA, Recherches en Linguistique: théorie des opérations énonciatives, Département de Recherches Linguistiques, Université de Paris VII.

Guillerault, M. and Laborde, C., 1982, Ambiguities in the description of a geometrical figure, in Language and language acquisition, p. 151-156, Edited by F. Lowenthal, F. Vandamme, J. Cordier. Plenum Press, London- New York.

Laborde, C., 1982, Language naturelle et écriture symbolique: deux codes en interaction dans l'enseignement mathématique, Thèse d'Etat, Université de Grenoble, IMAG.

Lowenthal, F., 1982, Contextual linguistics, Synthesis session, Future Projects in Language and Language acquisition, p. 365-371, Plenum Press, New York and London.

Rasolofoniaina, I, 1983, Conditions d'Apprentissage Mathématique par la lecture, Thème de 3ème cycle, Université de Strasbourg.

Walther, G., 1981, Autonomous learning and the reading of mathematical texts, Journal für Mathematik Didaktik Jahrgang 2 Heft 2, F. Schöningh, Paderborn.

THE OB-SERVING COMPUTER

F. Lowenthal and B. Harmegnies

Laboratoire NVCD
Departement de phonétique et psychoacoustique
University of Mons

1. INTRODUCTION

Lowenthal (1978) defined, some years ago, what is now known as "Non-verbal communication devices" (NVCD). A detailed definition and a long discussion concerning the relevance of the concept NVCD can be found in another paper (Lowenthal, this volume). Since 1978, much research concerning the use of NVCDs by several kinds of subjects, has been led in the NVCD lab at the University of Mons. This research shows the importance of this kind of approach (Lowenthal, 1980, 1984). A new research trend has been recently described (Harmegnies and Lowenthal, 1984) which will enable us to extend previous research and to study aspects of NVCDs which were not studied before. In order to reach our aim, we will use new means and we will place our investigations in a framework differing from the one previously used. On the one hand, we will use computerized NVCDs with children: the computer will give (or "serve") information; on the other hand we will focus on new research topics which could not be studied without having recourse to informatics: the computer will "observe" children.

The paper describes the device conceived to carry out our first experimental investigations within this new research trend. The results will be published later.

2. THE OBJECTS OF STUDY

In many research papers, authors are led to consider their subjects as black boxes. They can observe and describe only external manifestations, whilst they are mainly interested in what is happening inside. In fact, the essential problem is to determine how subjects manage to elaborate a response, given the kind and amount of information they received.

The most natural way to study the subjects consists in observing their productions, that is to say the black box's outputs. Lots of studies have dealt with this topic, which constitutes the main point of classical research into problem solving, mainly illustrated by NEWELL and SIMON (1972). In this kind of research, the first problem to deal with is the designing of materials which will lead the subject to manifest enough observable problem-solving behaviors.

NVCDs are very useful, from this point of view: by construction, they are full of built-in technical constraints and these constraints imply a gradual, step by step construction of the final product without interference from the observer. Each intermediate step can thus be the object of a specific study: it can be observed, measured and evaluated. Moreover, everything the subject does while he is trying to solve the problem restricts his possibilities of further action: the specificity of NVCDs thus enables the observer to better analyze and understand the subject's reasoning, simply by looking at the complete sequence of steps occurring in the construction of a final result. The observer can then use the logic suggested by the technical constraints together with this complete sequence of production steps and infer from these data the actual logical structure subjacent to the subject's production activity. Analyses of children's productions while they use a NVCD have already been developed to a great extent (Lowenthal, 1980, 1984, 1985). This kind of research will be further developed through the detailed analysis of videotaped documents: most of the experimental sessions were filmed and the children's verbal and non-verbal productions can thus easily be analyzed; one can find an example of such an analysis in Lowenthal and Saerens (this volume).

The result which can be obtained in this way are useful as far as the use of NVCDs in a teaching situation is concerned; these results also give very relevant information concerning clinical observations of children. Nevertheless, it is clear that this research trend is mainly focussed on the subjects' productions. Coming back to the Black Box analogy, one could state that only the output of the device is observed. But what about the input?

In the field of problem-solving study, the "input" of the subject is the information he requires in order to give the problem a final solution. Most of the information processing devices man has built have rather passive input procedures. For instance, the information a computer requires to perform some processing must be given by the user or the programmer. On the contrary, it can be said that man has very active input procedures which constitute, according to PIAGET (in Cellerier et al., 1978) one of the essential dimensions of the human mind. These procedures, commonly known as "perceptual activities" can thus certainly not be neglected. Several authors have noticed their importance. VANDAMME (1982) considers that production and perception are essentially different but equally important activities. STRAIGHT (1980) agrees with him: "the kinds of information processed by the comprehension mechanism are qualitatively different from those processed by the production mechanism".

Many researchers have already tried to analyze the ways in which subjects involved in a problem-solving activity get hold of the information they need. Most of them used devices meant to observe eye movements (Menu et al., 1978). Although this research trend gives very valuable information, it is nevertheless unsuited to the observation of perceptive strategies in young subjects; all the decices which are used introduce important constraints for the user: the subject is not free to move as he wishes, his cooperation is necessary, he must accept to "wear" an unconfortable and heavy tool during long sessions. HARMEGNIES (1979) concludes that all these reasons plead against the use of such devices with children. Moreover, all these devices are more or less biased; and YOUNG and SHEENA (1975) note that a reduction in the constraints imposed on the subject correspond to an augmentation of the bias. Furthermore, one cannot neglect LEVY-SCHOEN's remarks (1972): she observes that "visual fixation" and "perception" are not identical concepts: some details can be fixed by the subject without actually being perceived (Gainer and Obermayer, 1964; Mackworth et al., 1964), while the perception of a detail can occur without any eye fixation but by means of the peripheral retina (Antes and Edwards, 1973).

Given these restrictions, we decided to avoid the use of an eye-movement based technique and tried to build an original experimental setting. In the literature there are attempts to create such settings, without all the constraints imposed by the tools used to study eye movements. LEPLAT (1960) and KRIVOLHAVY (1965) studied perceptual activities without having recourse to eye movements. The basic principle of their research simply consists in the occultation of the information sources: the subject could then obtain the hidden information by an observable step (in KRIVOLHAVY's research all the information was hidden under several pieces of paper the subject had to raise one by one in order to obtain one piece of information at a time). We chose to adopt a similar device: NVCDs make it possible to observe how, step by step, a construction is produced; by analogy, we decided to create a setting where each elementary perceptive art, became observable.

Moreover, by using a single time counter for both data collection devices, the sequence of production acts and that of perception acts are both observable. Their interactions become thus, at least partially, observable.

3. RESEARCH HYPOTHESES

The main purpose of our experiments will be to check LOWENTHAL's hypotheses which assume that providing children with NVCDs, 1. helps to structure their perception, 2. acts as a starter for complex cognitive processes (Lowenthal, 1985). More precisely, the first question to be answered should be the NVCDs ability to organize the child's perceptive field. In order to prove this assumption, it would be necessary to emphasize specific structuration in the child's information acquisition routines. If a specific structuration is proved a second problem to solve would consist in discovering whether it is a good one. In other words, checking this hypothesis implies studying to what extent subjects familiar with NVCDs succeeded significantly better in getting the best information at the best moment. This question is obviously related to the elaboration of relations between the different elements of the perceptive field. One might wonder, on the one hand, whether users of NVCDs are able to build local relations between <u>some</u> of the relevant elements, or, on the other hand, whether they actually end up building global relations between <u>all</u> the relevant elements. Finally, we will try to determine to what extent behavioral adaptations can be observed and attributed to a specific NVCD effect.

In order to check these hypotheses, we have designed a specific experiment, which is described below.

R. THE EXPERIMENT

4.1. Experimental Design

Two group of subjects will be used in this experiment: one of "trained subjects", consisting in children familiar with NVCDs, and one of "untrained subjects", composed of children unfamiliar with NVCDs. All the subjects will be children in upper primary school classes.

The experimental treatment will be the same in both groups. First of all, the subject will be given two easy NVCD training problems, so that he has an opportunity to become familiar with the device used. Test problem I will be given immediately after this training session to the child. It will be immediately followed by test problem 2. These test problems are described in section 4.4.

4.2. Experimental Setting

We ask each subject to reproduce on a base-bord, using the bricks described by Cohors-Fresenborg (1978), a diagram which is "hidden" on a television screen controlled by a computer. The computer runs the presentation of pieces of this diagram at the request of the subject. In fact, the experimenter provides the subject with a television screen and a box furnished with 25 buttons. On the screen, the subject sees a big rectangle subdivided by a grid into 25 small rectangles. There is a reproduction of these rectangles and of this grid on top of the button box: there is a button in the centre of each small rectangle. The grid on the screen divides the diagram into twenty-five zones. Each pressure on a button provokes, in the corresponding zone of the screen, the apparition of the part of the diagram which the subject wishes to observe. This image remains present on the screen as long as the subject keeps pushing the corresponding button. Two images occupying two different zones cannot appear simultaneously on the screen.

4.3. Collection of Data

In order to record the production behaviors of our subjects, we chose to videotape all their problem-solving activities. This well-known technique is related to the framework of human ethology, illustrated, among others, by MONTAGNER (1978).

Another technique for controlling subjects productions consists in keeping track of them by means of a computer ☆. One could say that this is the "observing computer technique". We chose to use such a technique with an original aim: to study the subjects' information acquisition strategies. By doing this, we combined two roles of the computer. It would serve information at the request of the subject and, at the same time, observe the subject: that is the technique of the ob-serving computer.

Both data collecting techniques are combined by means of a single synchronizing device.

In fact, in our setting, whenever a zone of the diagram is observed, the computer identifies it and keeps track of the event and of the time counter value when the image appears and when the image disappears (the time counter was put on zero when the subject started working). A synchronizing signal generated by the computer is also recorded on the videotape. At the end of the execution, all the data concerning the perceptive activity are stocked on a disk and will be treated later.

☆ In this oral contribution to the 5th ICME, BJÖRKQVIST, (1984) presented such a recording technique: he uses a computer to study the productions of his subjects step by step.

4.4. More about the Task

4.4.1. Description of the NVCD used . We chose to confront the subjects with a well-known situation as far as our use of NVCDs is concerned: the subjects has to build a Dynamical Maze corresponding to a given diagram (Lowenthal, 1984) (Fig. 1). The material, invented by Cohors-Fresenborg, consists of plastic bricks which can be placed on as base-board. These bricks constitute the basic elements needed to create a "railway network": straight rails, curves, by-passes and switches. This material has built-in constraints which purposely restrict the number and the kind of combinations a child can make with the pieces: a "train" can only go through a "rail" in one direction, each piece is a "one way" element. There are several kinds of rails and mechanisms which function when a train goes through them: these mechanisms, the switches, enable a train to determine the direction which will be chosen by the next train. All these pieces represent a mechanical

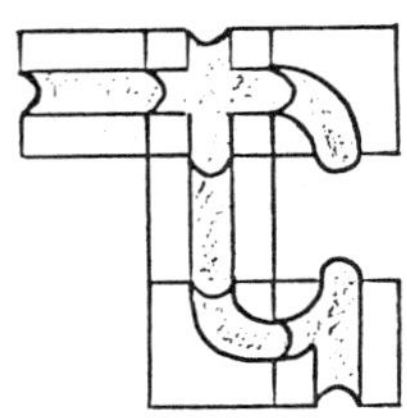

Fig. 1 The technical constraints make it impossible to place a piece; one should need a piece open at both ends and there is no piece like that.

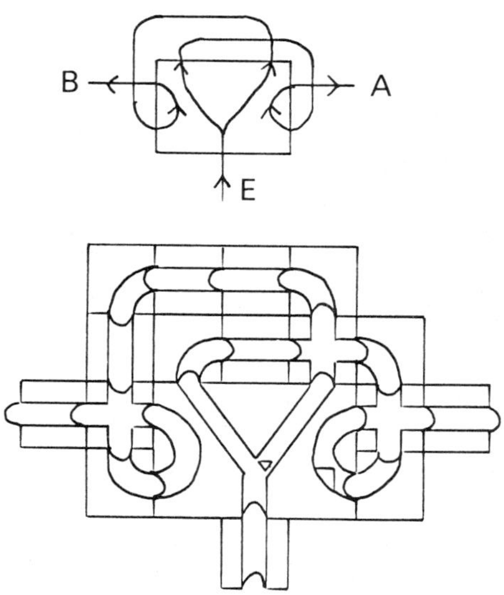

Fig.2 A diagram and the corresponding maze.

version of the hardware of a computer: the pieces can be combined and the mazes they form are finite automata where each orientation of the switches is an inner-state of the system. It is easy to ask children to build, actual mazes corresponding to a given diagram (as shown in figure 2) with the bricks placed at their disposal.

4.4.2. Test problems. The diagram corresponding to the first problem ("Test 1") is relatively simple and is shown in figure 3. Many zones are empty and the relevant parts of the information are concentrated on two vertical axes: the central one (zones 3, 8, 13, 18 and 23) and 4 zones of an adjacent axis (7, 12, 17, 22). The second problem ("Test 2") in more difficult: there are no empty

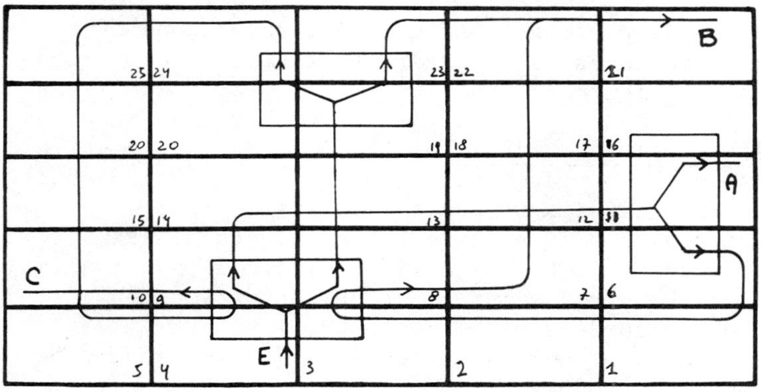

Fig. 3 The diagram of test problem 1.

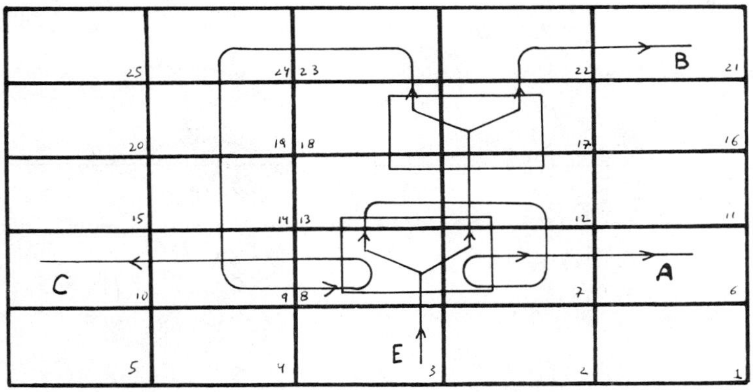

Fig. 4 The diagram of test problem 2.

zone and most, but not all, relevant parts of the information are concentrated on two vertical axes (zones 8, 13, 18, 23 and zones 7, 12, 17, 22) and one horizontal axis (zones 14, 13, 12, 11). All this can be seen in figure 4.

In order to treat the information obtained about the zones the subjects scanned, we chose to sort these zones according to the amount of information contained in them. We used the following criteria: a zone of type A is an empty zone; a zone of type B contains ambiguous information (e.g. the position of several pieces but not their orientation which must be deduced from information contained in other zones); a zone of type C is not empty but it contains non ambiguous information based on implicit indices (e.g. the "joint" shown on figure 4, zone 22 does not explicitly indicate the orientation of the "rails", but the nature of the NVCD used makes it possible to deduce this orientation); a zone of type D contains explicit clue which is not necessary to perform a correct construction (e.g. zone 21, figure 4); and a zone of type E contains an information which is necessary to perform a correct construction (e.g. zone 4, figure 4).

4.4.3. Assessment criteria. The final product of a child can easily be evaluated and compared with similar constructions built by other children. The criteria we use are the following:
a) the construction is a perfect copy of the diagram;
b) the network built by the child is logically equivalent to that shown on the diagram but it is not a perfect copy;
c) there are minor technical mistakes;
d) the subject's network is incomplete (i.e. it could be completed and become a production of type (a) or (b);
e) the subject's network is technically correct but not equivalent to that shown on the diagram;
f) the subject's network is neither technically correct nor equivalent to that shown on the diagram;
g) the subject made many technical mistakes and used a great number of incorrect bricks.

Table 1: analysis of the types of zones in TEST 1 and TEST 2

	TEST 1		TEST 2	
TYPE	NAME OF THE ZONE	NUMBER	NAME OF ZONE	NUMBER
A	1,2,4,5,15,16,20,25	8	--	0
B	9,14,19,22,23,24	6	1,2,5,7,10,12,13 14,15,16,17,20, 23,24,25	15
C	12,13	2	3,22	2
D	3,6,10,21	4	21	1
E	7,8,11,17,18	5	4,6,8,9,11,18,19	7

5. ANALYSIS OF RESULTS: PERSPECTIVES

One can use the information stocked by the computer only: this is enough to generate variables such as NO (the number of times a button was pressed), TT (total execution time), TO (total observation time), ROT (relative observation time), MOT (mean observation time), NO_i (number of times zone i was scanned), TO_i (total observation time of zone i),... All these variables can be used for global studies.

It is also possible to look at the sequence of perception acts in order to examine the type of strategy adopted by the subject: did he try to find a relevant element with an explicit index and did he then "follow the track"; did he search around a relevant element; did he look around a non relevant element; did he work by trial and error; did he go backwards (in the direction opposite to that shown by the indices) or did he completely ignore the directional indices? All these questions, and many others, can be answered with this experimental setting.

Other questions could be: did the subject look several times at adjacent zones in order to have a better idea of the situation (e.g. for the diagram shown on figure 3 many subjects scanned the sequence of zones: 8, 13, 12, 7 several times).

Moreover one can compare the sequence of observations and that of construction acts: did the subject look at zone 3 and immediately after put a brick in zone 24; did he check his construction, and how; what zones did he scan just before putting a wrong brick in zone 6; how did he react to the observation of zone 13;...?

There are thus many questions which can be answered; some of them are of a more global nature while others are of a more processual nature. Most of the variables generated by the measures made in this experimental setting can be used in both approaches.

6. DISCUSSION

The experimental setting described here can be used for several purposes; but does it help to study the mental processes involved when subjects use a NVCD? Can one use this setting to corroborate the hypotheses formulated? We want to show here that this is the case: this experimental setting is such that hypotheses can be formulated and tested.

A first question to be asked is: " Will 'trained' subjects identify non obvious visual clues faster than 'untrained subjects"? One can answer this question by using zones such as zone 22 in figure 4: no arrow can be seen on the diagram in that zone but the 'joint' clearly indicates that one must go from left to right and from bottom to top. Using the videotapes and the data stored by the computer, it is then possible to measure the amount of time each child needs to place his bricks correctly in that zone after scanning it for the first time. This might confirm the hypothesis that "NVCDs help subjects to organise their perceptive field".

Another question is: the subjects of the trained group need to scan zones 4, 9 and 8 as many times as "untrained subjects". In fact any one of these three zones is sufficient to determine unambigously the kind of switch which is used: one could thus measure, for each subject, the amount of time needed to choose the correct switch and place it correctly after one of the three zones has been scanned for the first time. One could also study without videotapes the total number of times these zones have been scanned (i.e. study a variable Z = NO4 + NO8 + NO9); one could also study a variable X = TO4 + TO8 + TO9 (i.e. X is the total observation time spent by the subject while scanning these three zones). The first approach is a processual one, while the two others are global; but these three approaches might confirm the hypothesis that "after structuring their perceptive field, users of NVCDs learn to discover the relevant elements".

One might also wonder whether "trained" subjects use a more efficient scanning strategy than "untrained" ones. Many things can be done to test this; some of them are very easy: how many times did a subject "leave the track" while scanning (e.g. scan zone 17 after zone 18 in figure 4); how many times did a subject scan back and forth (e.g. scan zone 13, then zone 18, then zone 13 again). Such results might confirm the hypothesis that "users of NVCDs build local relations between some of the relevant elements".

Interesting results could also be obtained by comparing the total observation time for each type of zone in both groups: this could give further information about the discovery of relevant elements. TO is a very useful variable: is it possible that "trained" subjects need less time to observe than "untrained" ones? Similar information concerning the construction could be obtained by studying TC = TT - TO (TC is te actual reflection and construction time which does not include any scanning period; one must note that some subjects keep pushing a button while placing bricks without looking at the image on the screen).

In order to check whether NVCDs users actually end up "building global relations between all the relevant elements", one could compute for each subject the number of times he scanned zone 20 (in figure 4) after he scanned zone 25 and zone 19 at least once; similar observations could be done fore zone 17, zone 2,... (in figure 4). The combination of all these results would give a variable which measures the number of unnecessary scannings: the study of this variable might show whether the subject is able to build global relations.

7. CONCLUSION

The experimental setting described in this paper is new: it uses a computer which gives or serves non-verbal information while keeping track of, or observing, the number and kind of non-verbal questions asked by the subject; this device is combined with a NVCD. This combination enables the experimenter to observe simultaneously the perception and production activities of the subject; moreover it enables him to observe how these activities seem to interact with one another and thus to find new results concerning the cognitive processes hidden in the "black box" (i.e. in the subject's mind). It is this new combination which enables the experimenter to formulate concrete hypotheses concerning the mental processes involved in the use of NVCDs. These hypotheses can now be tested: further papers will present and discuss the data drawn from our experiments.

8. REFERENCES

Antes, J.R., Edwards, D.C., 1973, Information processing in the visual periphery. Bulletin of the Psychonomic Society, 1, 351-253.

Björkqvist, O., Computers analysis of cognitive processes in problem solving (personal communication).

Cellerier , G., Papert, S., Voyat, G., 1968, Cybernétique et épistémologie, coll. Etudes d'épistémologie génétique (vol. 22), Paris, P.U.F.

Cohors-Fresenborg, E., 1978, Learning problem solving by developing automata networks, R.P.A., 46/47, 93-99.

Gainer, C.A., Obermayer, R.W., 1964, Pilot eye fixations while flying selected manoeuvers using two instrument panels. Human Factors 6, 485-501.

Harmegnies, B., 1979, L'étude des processus cognitifs via l'observation des mouvements oculaires; outils conceptuels et matériels (unpublished dissertation) Mons, UEMs.

Harmegnies, B., Lowenthal, F., 1984, Dispositifs de communication non verbale et ordinateurs, grkg/Humankybernetik, 25, 3, 115-124.

Krivolhavy, I., 1965, L'application des résultats de l'analyse algorithmique d'un système comprenant un élément humain à la construction des panneaux de réglage dans une usine électrique. Actes du IIIe. congrès de la S.E.L.F. Presses Universitaires de Bruxelles, 125-139.

Leplat, J., 1960, Exploration visuelle et système de repérage. Bull C.E.R.P., 9.

Levt-Schoen, A., 1972, Rapport entre mouvement des yeux et perception. In: Hecaen, H. (ed.): Neuro-psychologie de la perception visuelle, Paris, Masson, 72-92.

Lowenthal, F., 1978, Logic of natural language and games at primary school. R.P.A., 46/47, 133-140.

Lowenthal, F., 1980, Language learning and logic. In: Archehold, W.F.; Driver, R.H.; Orton, A. and Wood-Robinson, C. (eds.): Cognitive Development Research in Science and Mathematics, Leeds, University of Leeds, 121-128.
Lowenthal, F., Reflections about the relevance of non-verbal communication devices (in print in the proceedings of the third international conference on Language and Language Acquisition).
Lowenthal, F., 1984, Productions langagières d'enfants manipulant un dispositif non-verbal de communication, R.P.A., 69, 11-46.
Lowenthal, F., Non-verbal communication devices in language acquisition, R.P.A. (in print).
Mackworth, N.H., 1964, Eye movements during vigilance. Perceptual and motor skills, 18, 2, 397-402.
Menu , J., Tourneur, Y, Harmegnies, B., 1978, Les démarches exploratoires au cours de la résolution des items du test d'arrangement d'images. Document SEMME 781.015, Mons, UEMs.
Montagner, H., 1978, L'enfant et la communication, Paris, Pernod-Stock.
Newell, S., Simon, 1972, Human Problem Solving, Englewood Cliffs, Prentice-Hall.
Straight, S., 1980, Structural commonalities between comprehension and production, R.P.A., 55/56, 313-316.
Vandamme, F., 1982, Recognition and productions: two different skills in Language and Language Acquisitions, eds. Lowenthal, F., Vandamme, F., and Cordier, J.; Plenum Press, New York, 181-190.
Young, L., Sheena, D., 1975, Methods and designs, survey of eye movement recording methods. Behavior Research Methods and Experimentation 7,5, 197-429.

THE CHARACTER OF STUDENT KNOWLEDGE

Harry Osser

Faculty of Education
Queen's University
Kingston, Ontario, Canada

This chapter begins with an analysis of the critical differences between psychometric and clinical assessment practices with respect to their goals, assumptions and procedures. Then the topic of the complex nature of teacher judgment of students, and of observation as one type of clinical assessment, will be discussed. The argument continues with an examination of the several forms of knowledge that students employ in the course of their school performances and it concludes with an extended illustration of the use of clinical assessment in exploring student competence.

PSYCHOMETRIC AND CLINICAL ASSESSMENTS

Psychometric tests may be given to an individual or a group of students. Typically, the intention of the tester is to capture the product of learning, in order to answer the question, "what proportion of the school curriculum has been learned by this student, or these students?". The accepted procedure is for the tester to offer a standardized sequence of questions to those being assessed. The results of psychometric testing provide global information such as: "The student has an IQ of "X", is reading at a "Y" level, and is working at the "Z" level in mathematics". This information may be used to allocate a student to a particular group, or sub-group, in school.

In contrast, clinical assessments are individualized, both with respect to the one-to-one relationship between tester and student as well as to the form and content of the questions presented in the testing situation. Clinical assessment focusses on the question, "What does this student know?" and is not restricted to what the student has learned from the school curriculum. The shift in emphasis is from the identification of the products of school learning to the exploration of the processes involved in the generation of these products. The potential benefits of a clinical assessment include the provision of information about the individual student's difficulties with specific features of school work, such as failure to adhere to an appropriate sequence of "moves" in solving a mathematical problem. In other words, clinical tests are characteristically diagnostic , and refer to the variety of kinds of learning the student engages in.

The psychometric and clinical methods of assessment are related to distinctly different conceptual territories, so that beyond the obvious differences in means of assessment, there are also differences in goals. Underlying such differences are divergent preconceptions of the actual nature of students' school learning. Psychometric testing rests uppon a mechanical theory wich essentially represents the student simply as a passive absorber of lessons. This view results quite naturally in a testing instrument designed to measure the extent to which the student is able to reproduce the content of lessons. On the other hand, clinical testing stems from a theoretical position which emphasizes the autonomous character of the student's cognitive life. From this perspective, the student is seen as being actively engaged in organizing reality. The procedures used in clinical testing therefore represent attempts to discover how students are structuring, and restructuring, their school experiences, and what forms their knowledge takes.

The selection of a particular test procedure depends upon the nature of the judgment that has to be made. If initial screening is necessary to select the type and level of classroom for a student, a psychometric test might well be appropriate; such testing however will only provide information for classifying the student. If it is necessary to go beyond simple classification in order, for instance, to reveal, in a significant manner, details of the student's particular problems in responding to school work, clinical assessment procedures might be considered. Anyone who has attempted to make the student's cognitve processes explicit realizes that a variety of difficulties will be encountered in electing to use such a strategy. Yet the challenge must be taken up, as Rowntree (1977) puts it, "Rather than making the measureable important, we should make the important measureable" (p. 68).

THE COMPLEXITY AND LIMITATIONS OF TEACHERS' OBSERVATION OF STUDENTS

One of the common myths in education is that teachers have detailed knowledge of what parts of the curriculum individual students have mastered, what interests each possesses, what pleases and distresses them, and finally which teaching strategy must be used to harness the student's full intellectual and motivational resources in the processes of learning. This kind of extensive knowledge of the strengths and weaknesses of individual students in rare. One of the main reasons for this is that the basic question in evaluation, namely "What does this student know?", while appearing at first to be simple, turns out on analysis to be of considerable complexity. Answers to this question may be pursued in a number of different ways, including looking at students in classrooms. However, for observation to lead to a proper understanding of classroom events it is necessary to go beyond simple description. As Walker and Adelman (1975) put it, "The essence of observation is the creation of insight out of what might seem initially to be routine and commonplace (p. 18).

What often results from asking a teacher to describe a particular student is the provision of a verbal sketch wich includes such words as attentive, bored, underachiever, overachiever, slow learner, gifted, passive, hyperactive, clumsy, or well-coordinated. This list of descriptions represents only a small part of the total lexicon of educators. The function of these terms, or labels, is to permit the teacher to organize observational experience. Given that the number of adjectives typically available to an educated adult is very large, a pressing question is, "How does a teacher select specific terms in characterizing the school performance of a particular student?".

The argument so far is that teachers have sources of information about students, including observational data, out of which they develop a <u>category scheme</u>, consisting of a set of verbal descriptions, which they use to sort their students into various functional groups. Another basic question here is, "What particular kinds of information are critical constituents of the teacher's judgment?". To use a concrete example, how does a teacher come to characterize a student as, for example, a "good" reader? This judgment presumably derives from the teacher's observation that the student has most, or all, of the defining attributes of a good reader. Such a judgment derives from the teacher's possibly implicit theory of reading. One teacher, for example, might view a "good" reader as somebody who is flawless in pronouncing words gone over in a lesson. Another teacher might insist that the student should be able to read "new" words, for which no instructions have been directly given, before applying the label "good" reader.

THE TEACHER AS OBSERVER: IMPLICATIONS FOR STUDENT EVALUATION

It has been argued that a teacher develops a categorical scheme concerning student performance which helps in developing a profile of the strengths and weaknesses of any one student. Such a categorical scheme intersects with a teacher's implicit psychoeducational theories, and both place constraints on observation and evaluation. According to this viewpoint, observation is never pure and direct, but is instead filtered through the teacher's theories. These may take the form, for example, of beliefs that students from a particular family, or of a certain age, sex, race, or social class are capable (or incapable) of specific forms of academic development. Consider, for example, the real case of a boy who had been labelled a nonreader at 8 years of age. His teachers were pleased that under their guidance he had developed after four months of the new school year to the point where he could read simple books to them. They now judged him to be one year behind his classmates. In fact, through independent testing, it was discovered to the teachers' surprise, or even dismay, that the boy was now among the best in the class. His teachers' underlying assumption seems to have been that for a nonreader the most that one can reasonably expect, even with skilled teaching, is a small amount of progress in several months. This is an instance in which a theoretical viewpoint limited the teachers' evaluation, in fact resulting in a misevaluation of the student.

To answer the question, "What does this student know?" in any complete sense clearly demands more than simple observation. Consider these three examples of a student's performance in elementary mathematics:

Students work	Teacher's Judgment	Alternate Interpretation
119 +200 —— 319	Correct	Correct answer but poor strategy as student added from the left.
329 +852 —— 11711	Incorrect	Same strategy resulting in a wrong solution.
19 89 17 +16 —— 123	Incorrect	Righthand column added correctly but total of 31 was reversed, the student carried the 1.

these examples suggest that the mere observation of the end products of the student's work cannot, by itself, provide the teacher with anything like an adequate understanding of both what and how the student is learning. What is required is the use of procedures which permit the teacher to "observe" the student's learning processes. In these three examples, the explication of the problem-solving process was achieved by asking the student to say out aloud what he was doing as he was working at the problems. The next section provides an outline account of student competence and also a detailed illustration of how, by using clinical assessment procedures, a teacher, or a researcher, can arrive at useful conclusions about the strategies students use as they engage in school mathematics.

THE CONSTITUENTS OF SCHOOL KNOWLEDGE

In considering what the individual in the role of student has to know to be successful, an account of school work expressed solely in terms of the student's necessary grasp of the formal units of the curriculum would be inadequate: the student is equally exposed to the latent and informal curriculum of school values. One approach to the exploration of the character of school success is to determine the critical competencies that the student has to develop, and use, in order to cope with the demands of school. For example, Mehan (1980) has proposed that in order to participate effectively in the classroom students need to synchronize two forms of knowledge, one covering academic content, the other interactional form. In other words, it is not enough to have mastered the content of a subject-matter, the student must also be sensitive to the classroom rules that govern the presentation of such knowledge. Mehan's discussion on the nature of school knowledge is reformulated in the next section.

In developing a conceptual framework to make sense out of school experience, students, it will be argued, operate simultaneously with three kinds of knowledge. The first type is academic knowledge, or grasp of the content of subject-matter (Osser 1980). The second is social-cognitive knowledge which refers to the student's ability to both make meanings and intentions clear to others, and to understand their meanings and intentions (Osser, 1982). As Erickson and Shultz (1981) suggest:

> "The production of appropriate social behavior from moment to moment requires knowing what context one is in, and when contexts change, as well as knowing what behavior is considered appropriate in those contexts"
>
> (p. 147)

The third kind, metacognitve knowledge, relates to the student's skill in self-monitoring, illustrated by the use of feedback and the resultant corrective procedures. One linguistic example is the phenomenon of the "retraced false start", where the speaker detects a speech "error" and corrects it by the substitution, deletion, or addition of new verbal materials (MacWhinney and Osser, 1977). Other equally common examples are where the student might monitor task comprehension by asking such questions as, "What is this all about?", "Is it difficult?", "What is the next step?", and "Did I forget anything?".

A CLINICAL ASSESSMENT OF STUDENTS' METACOGNITIVE KNOWLEDGE

The following is an account of information generated by a study of a teacher and her students in a special class for children with substantial learning problems. One goal of the study was to analyze the role of different types of knowledge in children's mathematical performances. The following discussion will focus on metacognitive knowledge. The skills of metacognition in mathematical problem-solving include, for example, predicting, estimating and checking, that is they refer to the basic characteristics of thinking efficiently in learning situations (Flavell, 1979 and Brown, 1980). A second goal was to provide some information on the extent of the teacher's influence on students' performances. For example, the student might on some occasions operate with a "formulaic" strategy in school learning where some, or all, of the material would be memorized with minimal understanding. On other occasions the student might adopt a "generative" strategy. This represents an attempt to comprehend the relations between the elements of a classroom task. The student using a formulaic strategy may be voluntarily taking up, or be forced into, a dependent learning role, so that the student simply acts as a reproducer of the teacher's ideas, values, and problem-solving strategies; whereas the student using the generative strategy is more likely to have an independent learning role, and thus be an autonomous producer, or co-producer of knowledge with the teacher.

Clinical interviews of both the teacher and her students were employed to discover the extent to which a students take up work strategies as previously formulated by the teacher. In the following segment of an interview with the teacher, the focus is on the procedure of "checking" as an instance of monitoring school work. The interviewer speaks first and the teacher describes how the students are taught to check their work, and she comments on their fidelity in following her recommendations.

I: Do they check their work? Are they supposed to check their work?

T: I have taught most of them how to check their own subtraction questions by adding the bottom number with the answer to get the top one. Basically it's just a visual check. "Six divided by two. I have six put into groups of twos. I'm not going to get three, er twenty eight groups. I'm not going to get eighteen and I'm not going to get twelve. It doesn't make sense. I have to get a smaller number". Just usually checking to make sure if it makes sense or not. "If mother had five cookies and she gave four away, how many has she left? She can't possibly have nine".

I: You were saying that they do this or they don't do this?

T: They're encouraged to do it.

I: And to what extent do they try to do what?

T: I don't think that many of them do it. They simply get their work done and hand it in and so on to something else. Get some free time or whatever have you. It's simply "let's get it done". There are some that try to get their work right but they won't check it to see if they have them right or have answered all the questions. This happens quite often. I'll call them back and say "You didn't answer this question. You left that one out, you didn't do this one, what was the lesson?". "Oh, I didn't see those". So they didn't go back to see if they had everything down. "Did I have seven questions to copy off the board? Did I copy seven down? Did I copy five down?".

I: So there's two kinds of checking that they are not always doing. One is just to see that you've completed the actual work, that you've put answers down or copied all the problems down that you were given. And the other is to see whether you've done the work properly, when you actually did it.

T: Right. If you allow them they just don't do it. They think basically "I've got to get my work done and handed in, let the teacher check it over. Then if I get the work corrected, all right!.....".

The teacher's assumptions seem to be that students typically:

(1) Do not check to see whether they have copied all of the work from the blackboard, or whether they have answered every question.

(2) Know appropriate checking procedures which allow them to arrive at an estimate of the right answer, and thus can make progress toward it.

(3) Nevertheless the students typically leave it to her to check their work.

The following are samples from the interviews with two female students, P_1 (Terry, 8 years) and P_2 (Carol, 11 years):

I: Do you ever check your work?

P_1: No. I try to, but I just make the same math questions more worser, so I just leave them alone.

I: Oh, what do you mean that you make them worse?

P_1: Like if I hand'em in and I corrected them, right? Then she says "That one certainly isn't that". Say I handed that in and it was ten hundred, and she says "That wasn't it". I was wrong.

I: Do you think sometimes you have them right and then you change them when you check?

P_1: Well, I never check no more because I used to get them wrong. Now I sort of get them wrong the same way.

I: Do you ever check you work?

P_2: No I just look over it. Well when we're supposed to, like, um when Mrs W. was here we, we um did this kind of thing and she would put the answers up on the board and we would check them by ourselves and that's fun 'cause I like doing that.

I: Oh I see, you mean you would check to see if your answer was the same as hers?

P_2: They would show the real answer, the right answer, on the board. If you check over somebody else's work and you'd give, we did that last year and we passed over some. You take one person's work and you check over it to see if you got all your work right. You would take your own paper and check over.

Terry does not appear to share the values that the teacher attaches to checking, nor does she seem to understand the basic procedures for checking. She has apparently tried repeatedly to arrive at the "right" answer in the past without success; consequently she has given up on checking as a monitoring strategy. Carol, on the other hand, agrees that there is some value in checking, but her definition of it is very different from the teacher's: "checking" to Carol simply means comparing her answers to the teacher's "right" answer, or to other students' answers.

The teacher's view that the students have been taught and, therefore, must know how to check their math work is not validated by the evidence; however, the teacher's suggestion that her students expect her to check their work does receive some support. The teacher appears to underestimate the difficulties faced by her students when they are asked to check their math work. Succesful checking (or monitoring) presupposes: (1) that the student is competent in basic mathematical operations, such that errors can be detected; and (2) that knowledge of correction procedures, including estimation and prediction, are available. These competencies even when developed by "learning disabled" children are often inaccessible, as their diminished confidence in their stock of knowledge results in the adoption of the strategy of "playing it safe", exhibited in Terry's abandonment of her checking procedures, and Carol's delight in using other's presumably "right" answers. The two students seem to be operating, at least in the math class, with a general formulaic strategy in learning.

INDIVIDUAL DIFFERENCES AND SCHOOL COMPETENCE

Differences among pupils in academic performance can be understood by referring not only to variations in academic and social-cognitive knowledge as Mehan (1980) suggests, but by considering possible significant effects of differences in metacognitive knowledge. To return to the topic of the monitoring of comprehension, it is conceivable that students who closely monitor their understanding of a mathematical task may select a different and more appropriate problem-solving approach if and when it appears that they are not making progress. On the other hand, students who monitor their performances in a casual manner may miss the clues that indicate they are not on the right track, so that faced with difficulties they cannot precisely define they might be inclined to give up, as Terry did, rather than to persist with the work.

It is conceivable that a student who typically employs a "close monitoring" strategy for a given set of school problems will be more likely to adopt the generative mode of learning and its associated student role of producer, or co-producer, of knowledge. The corresponding hypothesis is that a student who typically adheres to the "casual monitoring" strategy for a given set of school problems will adopt the formulaic mode of learning and with it the student role of reproducer of knowledge. The performances of Terry and Carol seem to fit this latter characterization. However, it is likely that <u>every</u> student will follow a "close monitoring" strategy for some school problems and a "casual monitoring" strategy for others. If this is the case the student, if observed for a long enough time period, will likely vacillate from independence of the teacher to dependence on her as a function, among other things, of the subject-matter being taught and the specific context of learning, including the quality of personal relationships in the classroom.

One clear suggestion from the results of part of the clinical assessment of Terry an Carol is that they conferred their own meaning on school experience which did not coincide with what had been proposed by the teacher. This conception of students as shapers and interpreters of experience receives general support from the psychological theory of human action espoused by Von Cranach (1982). From such a theory it is possible to predict that students would inevitably develop alternative conceptions of "school mathematics", or any other part of the formal and informal curriculum, which would be quite distinct from the "authorized" versions. What the student's "alternative conception or framework" may be is however problematic. We are very far from having an adequate account of the student's interpretation of school work. The clinical assessment procedure appears to provide a useful entry to the student's world.

REFERENCES

Brown, A., 1980, Metacognitive development and reading. In R.L. Spiro, et al., Theoretical issues in reading comprehension. New York: Erlbaum Associates.

Erickson, F., and Shultz, J., 1981, When is a context? Some issues and methods in the analysis of social competence. In: J. Green and C. Wallat, eds., Ethnography and language in educational settings. Norwood, N.J.: Ablex Publishing, 147-160.

Flavell, J.H., 1979, Metacognition and cognitive monitoring: a new area of cognitive-developmental enquiry. American Psychologist, 34, 10.

MacWhinney, B., and Osser, H., 1977, Verbal planning functions in children's speech. Child Development, 48, 978-985.

Mehan, H., 1980, The competent student. Anthropology and Education Quarterly, 11, 3, 131-152.

Osser, H., 1980, The structure of children's school knowledge. Revue de Phonetique Appliquée, 55-56, 231-242.

Osser H., 1982, The child's construction of the social order of the classroom. In F. Lowenthal, F. Van Damme and B. Cordier, eds., Language Acquisition. New York: Plenum Press.

Rowentree, D., 1977, Assessing Students. London, Harper and Row.

Von Cranach, M., 1982, The psychological study of goal-directed action: basic issues. In M. Von Cranach an R. Harre, eds., The Analysis of Action. Cambridge: Cambridge University Press, 35-73.

Walker, R., and Adelman, C., 1975, A guide to Classroom Observation. London: Methuen.

RULES IN ARITHMETIC. LEARNING THE BASIC FACTS

Miriam A.D. Wolters

Developmental Psychology Department

University of Utrecht, Holland

1. INTRODUCTION

At the present time there is no teacher who can actually say that all is well with the teaching of arithmetic.

There are far too many children who dislike arithmetic or worse, children who think it is a 'stupid' schoolsubject. With relatively few exceptions, this situation is quite general and has come to be taken for granted.

The major part in arithmetic of both set homework and of class-work is devoted to the acquisition of techniques and to the improvement of skill in computations. There is, of course, a difference between understanding the working of a technique and understanding its subject matter. One needs only to ask a child to do an operation and he will give you the right answer. It is very easy for a teacher to think that this child understands arithmetic when in fact he might not. This child may well be conversant with all the technicalities of operations without having much idea about what sort of thing an operation is.

We know from pupils' behaviour that the teaching of understanding in arithmetic is not very successful. Although the younger pupils may find a certain delight in the formality of the processes at the moment they are taught, this soon wears off, and the popularity of arithmetic declines as the pupils grow older. They may even get very bored. The memory load increases beyond what can be easily borne. In spite of all the efforts of individual teachers, pupils' attitudes in many cases are still plainly negative, or if there is a positive attitude it is more often due to a sense of achievement at having a higher mark than somebody else.

2. THE RELATION BETWEEN ARITHMETIC AND ALGEBRA

Let us look first at what it seems reasonable to cover in a school course of arithmetic. I assume we all agree that the four operations: addition, subtraction, multiplication and division should be learnt, as applied to whole numbers and fractions (the latter expressed both in fractions and in decimals). The concepts related to these computations are those dealing with numbers as such, and with the properties of numbers, as well as the properties of operations that can be performed on numbers. There is no sense nowadays in making a distinction between arithmetic and algebra. There is so much connection between them that it is impossible to speak about one without introducing the other. It may sound revolutionary that certain algebraical notions should be known before arithmetical operations can be mastered. However it is not. Teachers who are aware of pupils' learning requirements have always attempted to clarify the underlying concepts before introducing a new process. What I want to suggest is that the clarifications used by these teachers, on the whole implicitly, should be made explicit and be organised in a curriculum. One of the things we must not forget is that pupils will not spontaneously acquire algebraical concepts. And this is simply because everyday-life does not provide children with experiences which will enable them to grasp them. This cannot be stressed too often.

Let me give you an example. For a pupil to do arithmetical operations it is very desirable that he is familiar with the commutative law. Many pupils, even at the age of 9, appear to be quite unaware of it; or if they are, they regard it in isolation, as a sort of curiosity rather than as a property of an operation.

It is the teacher's task, therefore, to provide situations, sometimes artificially in the classroom, so that the pupils are forced to pay attention to the commutative law. Much of the work of Daj (1972, 1976) has been spent on teaching algebra as a basis for teaching arithmetic.

3. LEARNING THE BASIC FACTS IN ADDITION AND SUBTRACTION

3.1. The Use of Strategies

A knowledge of the basic facts in addition and subtraction is essential for success in elementary school mathematics. Moreover, knowledge of the facts of each operation is necessary before presenting the algorithm for that operation. It happens, time and again, that children are frustrated by an inadequate mastery of the facts. If one looks for the reasons of their failures it appears that these children fail to use strategies that can help them remember the facts. 'Successful' children often use easy facts to find answers to harder ones, e.g. 6 + 6 = 12, so 6 + 7 is one more, 13 (Myers and Thornton 1977). Houlihan and Ginsburg (1981) argue that children's strategies are sometimes partly based on what is taught in school and partly based on their own inventiveness. Children assimilate what is taught in school to what they already know and the result is an 'invented strategy'. Rathmell (1978), working with young primary school children, found a positive correlation between high achievement scores on addition fact tests and the use of strategies for solving additions. Strategies used by the pupils in his study are such as counting, doubling, thinking one more or one less than a known fact, using ten and recognizing the commutative law. In 1979 he argues that more capable pupils often discover and use many different strategies, but average and low-ability pupils are less likely to discover these strategies by themselves.

Consequently, explicit instruction for thinking strategies will be necessary if children are to learn more mature and efficient methods for solving facts. The question is however, what kind of strategies are effective for helping children in learning basic facts.

3.1.1. The commutativity principle. Recently, research has been done on children's use of the commutativity principle as a strategy (Ginsburg, 1982; Baroody et al., 1983). Some of these results suggest that children up to 9 years may not be aware of opportunities to use the commutative idea. Ginsburg (1982) for example notes that elementary school children may appreciate the principle with very small numbers (e.g. 3 + 2 and 2 + 3) but not necessarily with larger ones. On the other hand, in Baroody's study (1983), 59 % of the first graders were consistent users of the principle to shortcut computational effort. This may suggest that the principle is well-known to young children. Perhaps commutativity is such a highly salient short-cut, because this property may be easily verified by informal experiences with arithmetic. For example, when counting with concrete material, young children do not consistently respect addend order as they count up the addends (Baroody et al. 1983; Carpenter & Moser, 1982). On the other hand first-grade children who do not have a rich informal experience with arithmetic infrequently use commutativity as a shortcut (Baroody et al., 1983). This evidence suggests the utility of teaching first-graders a strategy that makes the commutativity principle explicit. Although some pupils use this principle intuitively, many first-graders may not.

3.1.2. The addition-subtraction complement. A principle used for helping to memorize basic facts in subtraction is the addition-subtraction complement (Baroody et al., 1983). This principle is based on addition and subtraction as complementary processes (e.g. 5 + 4 = 9, 9 - 4 = 5). Nearly all of the third-graders (83 %) in the Baroody study used the complement principle in a majority of the subtractions and were very consistent in doing so. Only 39 % of the first- and second- graders used this principle.

3.1.3. The n+ 1 progressing principle. The use of the n + 1 progression principle as a strategy is examined in many studies (Baroody et al. 1983; Carnine & Stein, 1981; Myers & Thornton, 1977; Thornton, 1978). The n + 1 progression principles should be regarded as a set of principles including doubles (e.g. 4 + 4, 5 + 5), doubles plus 1 (e.g. 4 + 5, 6 + 7), sharing numbers to make a double (e.g. 6 + 8, 9 + 7).

In all these studies the authors are very optimistic about the effectiveness of the use of those principles. In the Baroody study however the use of the n + 1 principles was lower than was expected. In particular the third-graders, who might have been expected to use the principles most, seldom used them. None of the third-graders, 33 % of the second-graders and only 22 % of the first-graders used the principle at the first opportunity. This principle does not seem to be as salient as the commutativity principle. Its use is of course greatly facilitated by the presentation of orderly sequences of sums e.g. 2 + 1 = 3
2 + 2 = 4
2 + 3 = 5 etc.
But such a sequential arrangement of sums is unlikely to occur, except in the school context. This might be one of the reasons for the inconsistent and relatively infrequent use of the principle.

However, there might be another reason. The n + 1 progression principles are very useful for learning the basic facts in addition. Children might try to use it for basic facts in subtraction as well. At first sight this seems to be a good idea. Let me give you an example: 8 - 4 = 4, 9 - 4 = 5 and 7 - 4 = 3. In this example the one more (9-4) and the one less (7 - 4) concept works. When you use the same concepts for the second subtrahend: 8 - 5 and 8 - 3 it does not work anymore. The one more concept (8 - 5) gives you instead of one more, one less (8 - 5 = 3); and the one less (8 - 3) gives you one more (8 - 3 = 5). The use of sharing numbers to make a double which is very efficient for addition tables, is very difficult for subtractions. In the example of 6 + 8 to think of doubles (7 + 7) is of course a good strategy, but to think of 8 - 4 in the example of 7 - 5 gives you the wrong answer. Some anecdotical evidence [1] suggests that especially the slow and average learners

---

[1]. Quadvlieg T., Personal communication Psychological Laboratory Utrecht 1983.

try to use the n + 1 progression principles in subtraction. Moreover, it is very difficult to explain to these children why these principles do not always work in subtraction. On the other hand when children try to use the commutativity principle in subtraction it is very easy to explain and show them why it does not work. By presenting an example in symbols or in cubes they intuitively grasp the falseness of the idea (e.g. 8 - 4, 4 - 8).

3.2. Addition versus Subtraction facts

In general the basic facts in subtraction are harder to learn and memorize than the basic facts in addition. Baroody et al (1983) found that the mean accuracy with subtraction problems (12 - 6, 18 - 9, 20 - 10, 22 - 11) was lower for all three grade levels: first-graders, 43 % with subtractions and 78 % with additions; second-graders, 61 % against 90 %; third-graders, 92 % against 96 %. Even after an eight-week instruction period there is a significant difference in mean accuracy with addition and subtraction facts (fig. 1, Thornton 1978).

The instruction for the second-graders was based on a special sequence in order to emphasize thinking strategies for remembering the facts. The strategies emphasized during the instruction period were mainly the n + 1 set of principles mentioned above. Those who mastered the addition facts in the eight-week period turned to subtraction facts and were prompted to "think of the addition fact" to find their answers.

An example: 15 - 8, the pupil has to think of 8 + 7 = 15, so 15 - 8 = 7 (Meyers and Thornton 1977).

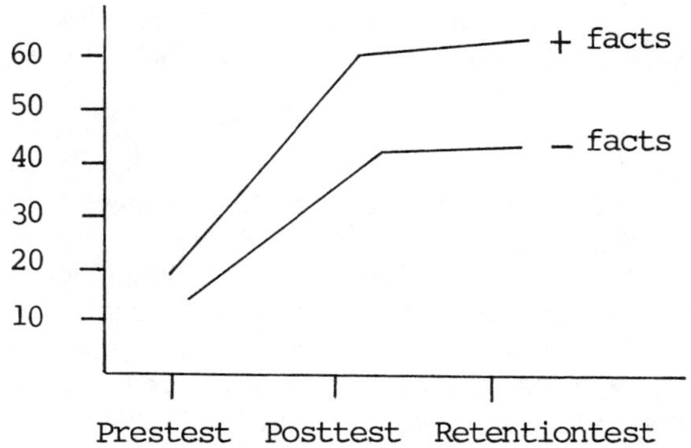

Fig. 1. Mean accuracy scores on pre- post- and retention tests for facts in addition and subtraction (Thornton 1978).

4. TEACHING THE BASIC FACTS IN ADDITION AND SUBTRACTION

Let us now return to the question posed earlier on: what kind of strategies are effective in helping children to learn basic facts?

To find an answer one might look at the strategies successful children use to learn basic facts. There is a change however that not all of these strategies are also effective for the average and low-ability pupils. This might be the case for instance for the n + 1 progression principle mentioned before. Another way to find an answer to the question is to look at the content or structure of mathematics related to the basic facts. Basic facts in addition and subtraction can be defined mathematically as the two computations: addition and subtraction.

A basic fact in addition and subtraction is a true equation formed by any grouping of a pair of the ten digits. If you count them there are 132 basic facts: 66 in addition and the same amount (66) in subtraction. We have excluded the facts in addition having sums greater than 10.

According to current practice all 132 basic facts have to be mastered at the end of grade 1. Practice or drill is an essential part of the arithmetic program to achieve this goal. To achieve mastery of a fact a pupil must be able to respond to it spontaneously and with assurance. Practice may consist largely of a series of repetitions over a period of time of answers to the sum of two digits. We do not recommend a program of this kind. The learning that is taking place is what we call learning by rote. On the one hand this type of learning is not very effective and on the other it is very time-consuming.

In our project 'Algebra in the elementary school' (OPEREL) we are trying to reduce the number of basic facts to be learned by rote. The principle behind this part of the program is that most of the basic facts should be derived primarily from the known facts by the application of properties of operations.

Fig. 2. gives the 66 basic facts in addition. This table should be read as follows: in order to find a basic fact, e.g. 5 + 3 locate the column with the number 5 and the row with the number 3. At the intersection of both you will find the letter R. The R means that this basic fact has to be learned by rote. As you can see in fig. 2. there are altogether 25 basic facts to be learned by rote. However, 9 out of these 25 are facts to which only 1 has to be added. They are easy to remember because adding 1 to a number gives the next number. So in fact there are 16 basic facts pupils have to memorize. All the other basic facts, 107 in all, should be derived from the 25 memorized by the application of

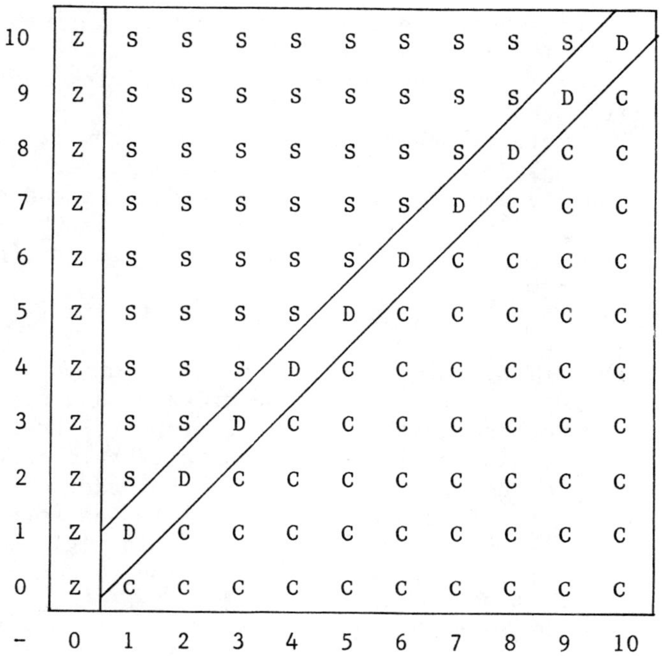

Fig. 2. Rules for the basic facts in addition.

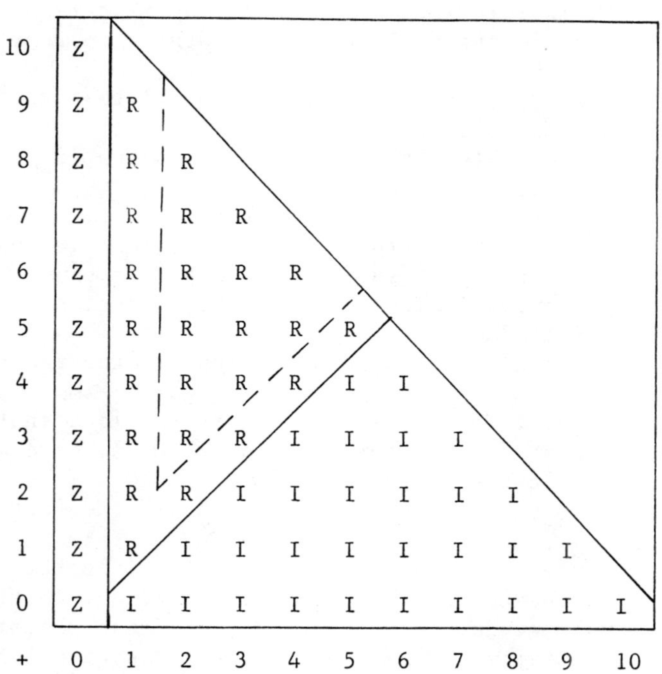

Fig. 3. Rules for the basic facts in subtraction.

properties of operations. The pupils derive the 107 'new' facts by applying 'rules'. These rules are given names and are based on the properties of operations. The two rules for deriving 'new' facts in addition are:

-zero rule. This rule is based on the zero property of addition. It simply states: adding zero to a number leaves the quantity unchanged. In other words: the sum of a number plus 0 is the same as that number (0 is the identity element for addition). In formal language: for every a, $a + 0 = a$.
This rule has to be used with all the facts indicated by Z in fig. 2.

-interchange rule. This rule is based on the commutative property of addition. It states: the order of adding two numbers does not change the sum. This property can be stated more concisely with the use of algebraic language. For every a and b, $a + b = b + a$. (The order of the addends does not affect the sum). This rule has to be used with all the facts indicated by I in fig. 2.

Fig. 3. gives the 66 basic facts in subtraction.

There are no basic facts in subtraction which have to be learned by rote.

The four rules for deriving the basic facts in subtraction are:

- zero rule. This rule is based on the idea that zero is a right-hand identity element for subtraction. For every a, $a - 0 = a$.
This rule has to be used with all the facts indicated by Z in fig. 3.

-cannot rule. This rule is based on the idea that it is only possible to subtract in examples of the $a - b = c$ when $a \geqslant b$, when a is equal to or greater than b.
Another way of formulating this rule is based on the principle that subtraction does not have the commutative property. A single example is sufficient to prove this point: $3 - 1$ is not equal to $1 - 3$.
This rule has to be used with all the facts indicated by C in fig. 3.

-disappearing rule. This rule is based on the idea that every number has an inverse for addition. The sum of a number and its inverse is zero (0): $a - a = 0$.
This rule has to be used with all the facts indicated by D.

-subtraction rule. This rule is based on the idea that subtraction is the inverse operation of addition. In teaching this rule we use decomposition exercises of numbers up to 10. An example $5 \swarrow^{9} \searrow 4$. The number 9 can be decomposed in the two numbers 5 and 4. One can read this diagram in different ways. $5 + 4 = 9$ $(4 + 5 = 9)$ but also $9 - 5 = 4$

(and 9 - 4 = 5). In general: ${}_{a}{}^{c}{}_{b}$ c - a = b. This rule has to be used with all the facts indicated by S.

5. CONCLUSION

In conclusion I would like to go back to the questions I mentioned before. Firstly, what kind of strategies are effective in helping children to learn basic facts? We have seen that we have to be very careful with teaching strategies used by bright children to average and low-ability pupils. Some of these strategies are not generally applicable. An example is the use of the n + 1 progression principle as a strategy. As we have seen this principle is an effective strategy for addition facts but it is not a strategy one can use for subtraction facts.

Secondly, learning basic facts by rote is not very effective and very time-consuming.

As far as effectiveness is concerned I regret that I do not have as yet all the data to prove this point. However as far as the time is concerned I can say a little more. It is clear from our experiments in the schools participating in this project that the pupils need about six to eight weeks to master the basic facts in addition and subtraction by applying the rules just stated. In the present situation the time required to learn the basic facts is approximately six months. In addition there is another advantage, namely, that these rules can be used for addition and subtraction with larger numbers.

REFERENCES

Baroody, A.J., Ginsburg, H.P., Waxman, B., 1983, Children's use of mathematical structure. Journal for Research in Mathematics Education, 14 (3), 156-168.

Carnine, D.W. and Stein, M., 1982, Organizational Strategies and Practice Procedures for Teaching Basic Facts. Journal for Research in Mathematics Education. Jan., 65-69.

Dienes, Z.P., 1960, Building up mathematics. London.

Grossnickle, F.E. and Reckzak, J., 1973, Discovering meanings in elementary school mathematics. 6th edition. Holt, Rinehart and Winston, Inc.

Ho Ngok Daj; 1982, De psychologische problemen van het moderne wiskunde-onderwijs in de eerste klassen van de basisschool (1976). In M. Wolters (red.), Perspectieven op ontwikkelend onderwijs. Vertalingen-bundel Russische Psychologie. R.U. Utrecht, IPAW.

Ho Ngok Daj, 1972, Kunnen tweedeklassers algebraïsche operaties leren uitvoeren? Voprosy psichologii (1) 85-97 (in het russ.).

Houlihan, D.M., and Ginsburg, H.P., 1981, The addition methods of first- and second-grade children. Journal for Research in Mathematics Education. March, 95-106.

Myers, A.C., and Thornton C.A., 1977, The learning disabled child. Learning the basic facts. The Arithmetic Teacher, 25 (3), 46-50.

Rathmell, E.C., 1978, Using thinking strategies to learn the basic facts. In M. Suydam, ed., Developing computational skills. The 1978 Yearbook of the National Council of Teachers of Mathematics. NCTM.

Rathmell, E.C., 1979, A reply to 'Formal thinking strategies: a prerequisite for learning basic facts?' Journal for Research in Mathematics Education. Nov., 374-377.

Thornton, C.A., 1978, Emphasizing thinking strategies in basic fact instruction. Journal for Research in Mathematics Education. May, 214-227.

Wolters, M., Perrenet, J. en Kwakman, F., 1981, Project Algebra op de basisschool. Interimrapport II, R.U. Utrecht, IPAW.

Wolters, M., 1978, Van Rekenen naar Algebra. Een ontwikkelingspsychologische analyse. Dissertatie R.U. Utrecht.

Wolters, M., 1983, Rules in arithmetic. In verbal or symbolic form? Paper presented at the 3rd language and language acquisition conference "Pragmatics and education" 21-25 March. Gent.

LANGUAGE DISORDERS

THE COMMUNICATIVE IMPACT OF NON-FLUENT APHASIA ON THE DIALOG BEHAVIOR OF LINGUISTICALLY UNIMPAIRED PARTNERS

Ria De Bleser
Research Group for Aphasia and Cognitive Disturbances
Heidrun Weisman
Depatment of Neurology
RWTH Aachen, West Germany

ABSTRACT

This paper reports on the specific linguistic and communicative strategies cooperative non-aphasic partners spontaneously employ in negotiation dialogs with non-fluent aphasics (Broca's and recovered global aphasics).

The following hypotheses were examined:

1. The non-fluent aphasic's dialog behaviour is deviant. His communicative disorder depends mainly on the disturbed expressive language abilities, which make it difficult for him to assert himself. As a consequence, his strategies of verbal negotiation are affected.

2. The dialog behaviour of non-aphasic patients changes when they communicate with aphasic partners.

3. The communicative adaption of non-aphasic patients is apparent in the dialog strategies used . In comparison with dialog strategies used between two non-aphasic partners, the non-aphasic patient uses more dialog remediation strategies of a specifically linguistic nature when his partner is aphasic.

Hypothesis 2 was confirmed, since a significant difference was found between the non-aphasics' use of dialog categories when communicating with other non-aphasics and in their dialogss with aphasic patients.

They used certain language specific categories exclusively in their dialogs with aphasics. These were the language categories which had a clear repair function in the dialog and could provide direct feedback.

One mixed category was also used significantly more often by non-aphasics in the constellation with an aphasic patient. In this category belong units which ask whether the partner's intention has been understood correctly. Thus, indirect feedback is provided. This confirms and refines hypothesis 3.

There was no evidence to support hypothesis 1 that the aphasic patient is disturbed in his dialog behaviour. This can certainly be explained by the cooperative attitude of the non-aphasic partner, not only as shown in his use of feedback categories, but also in the reduced syntactic complexity of his utterances. Thus, the 'normality' of the aphasic's dialog behaviour is only apparent, as it hinges on the naively remedial attitude of the non-aphasic dialog partner. Such cooperative attitude of the partner does not necessarily reflect the real life situation of aphasic patients.

1. INTRODUCTION

Aphasia research starts from the observation that elementary communicative abilities are selectively spared with left hemisphere damage. As a result, aphasia is characteristically described in terms of a disturbance of the instrumentalities of language, and not of the condition of its use.

All four standard aphasic syndromes (global, Wernicke's, Broca's and anomic aphasia) show disturbances of all linguistic components (syntax, semantics, and phonology) in all modalities (reading, writing, language production and comprehension). The exact nature and extent of these disturbances determine the characteristic symptoms for each syndrome.

It has been experimentally demonstrated by Boller and Green (1972) that even patients with global aphasia and severe Wernicke's aphasia - who have hardly any language comprehension left - reacted differently to meaningless material than to meaningful stimuli. They gave significantly more quizzical responses to semantic jargon (e.g. "How would you bay the hair of a querulous?"), and especially to phonemic jargon (e.g. "The mome raths outgrabe") and sentences of a foreign language (in this case French) than they did to English stiluli.

The preserved pragmatic competence of left hemisphere damaged aphasic patients, in particular of those with relatively good comprehension (Broca's aphasia), has also been investigated in comparison to right hemisphere damaged non-aphasic patients in

Table 1. Characteristic symptoms for the four standard aphasic syndromes

	Anomic aphasia	Wernicke's aphasia	Broca's aphasia	Global aphasia
Language production	mostly fluent	fluent	non-fluent	sparse or nihil, also speech automatisms
Articulation	mostly unimpaired	mostly unimpaired	often dysarthric	mostly dysarthric
Prosody	mostly unimpaired	mostly unimpaired	often monotonous	often monotonous, exept with speech automatisms
Syntax	hardly impaired; fragmentary sentences	Paragrammatism (doubling and blending of phrases)	Agrammatism (only simple sentences, lacking function words)	only single words, speech automatisms, stereotypes
Semantics	compensatory strategies with word finding problems; semantic paraphasias	many semantic paraphasias, often wild semantic neologisms, in severest form sesemantic jargon	rather limited vocabulary, few semantic paraphasias	very limited vocabulary, wild paraphasias
Phonology	some phonemic paraphasias	many phonemic paraphasias, also neologisms, phonemic jargon	many phonemic paraphasias	very many phonemic paraphasias and neologisms
Comprehension	mildly impaired	very impaired	mildly impaired	very impaired

several recent studies. The latter have shown deviations in the interpretation of humor (Brownell et al., 1983), in drawing inferences between two sentences (Goodenough et al., 1982), and in the processing of texts (Wapner et al., 1981). These and other experimental results have led to the assumption of a double dissociation between left and right hemisphere disturbances (Zurif, 1980), with disturbed language and preserved pragmatics in left hemisphere aphasic patients, preserved language and disturbed pragmatics in right hemisphere non-aphasics.

Rather than concentrating on the disturbed linguistic and preserved pragmatic abilities of the aphasic patient himself, this paper reports on the changes a linguistically impaired partner induces in the verbal communicative behaviour of a linguistically unimpaired dialog partner. Dialogs involving negotiation were chosen as a basis of observation, because their successful course towards a solution involves compromises acceptable to both partners. This requires that the partners truly interact, and that each participant is able to express his own point of view and convince the other by verbal argumentation. If only one of the partners can assert himself, the weight of the dialog will fall predominantly on his side and the dialog course will be disturbed. If the other partner is limited to making concessions, he will eventually be forced into withdrawal and the dialog will come to a premature end. This would run counter to the communicative purpose of both partners in a cooperative situation, namely, to maintain a dialog and steer it towards a negotiated compromise.

The aim of this study was to discover and describe the repair mechanisms spontaneously used by linguistically unimpaired subjects in order to counterbalance the disturbing factors introduced into the dialog by the aphasic patient. This study can be of interest to the general issue of verbal interaction in "noisy" communicative situations. It should further be a practical relevance for the elaboration of a realistic aphasia therapy approach (De Bleser and Weismann, 1981) and for the counselling of families with aphasic relatives.

2. SUBJECTS

Two patients with Broca's aphasia and two non-brain-damaged patients without aphasia were selected from the patient population of the Department of Neurology, RWTH Aachen, West-Germany. They were matched for sex (1 male and 1 female each), age (Women between 25 and 30, man between 40 and 45) and socio-economic status (middle-class; professional men, non-professional women).

Patients with Broca's aphasia were chosen rather than other aphasic syndromes because their relatively good language comprehension allows them to take up their hearer roles in a dialog, which patients with global or Wernicke's aphasia cannot do. However, in contrast to patients with anomic aphasia, the expressive language of the Broca patient is severely disturbed, which prevents an unimpaired partner from fulfilling his hearer role normally. Although the propositional content of the Broca's patient's speech is largely preserved (Ulatowska, 1983), it is difficult for a partner to understand the message he intends to convey: frequent phonemic paraphasias (e.g. woban for woman), agrammatic syntax (short, simple sentences, often lacking function words), and speech and language effort distort the expressive language severely. If the unimpaired partner is not cooperative, he can exploit his undisturbed language abilities to take over the course of the dialog, especially since his aphasic partner does not have the expressive linguistic means necessary for assertive argumentation. If, however, he his cooperative, he may use his linguistic superiority to provide mechanisms functioning as dialog repairs. The non-aphasic patients of this study were embedded in a cooperative setting. They had no therapeutic or didactic experience, which would have caused protective and directive, rather than neutral cooperative intervention. Although they were told that the patients had difficulties with language, they had not had any previous personal experience with aphasic patients and were not given any further details on the exact nature of the aphasic disturbances.

3. MATERIAL

3.1. Construction of Model Dialogs

Three balanced negotiation dialogs were used as a model. The scripts were constructed according to predefined content, structural, and pragmatic criteria. The point of departure was always a conflict between the dialog partners, the target of the dialog being the negotiation of an acceptable compromise.

The following topics were chosen:

D.1. Vegetable soup
The wife wants to make a soup, her husband would rather go out to eat.

D.2: Trading in
The customer in the shop wants his money back without a sales slip, the sales clerk refuses.

D.3: At the customs
The woman crossing the border has too many goods and refuses to pay. The customs officer insists that prescriptions be followed.

The topics reflect different pragmatic features of hierarchy in partner relationship, of social situation, and of speech register (see Henne and Rehbock, 1979). The hierarchical features of a partner relation can be symmetrical (equivalent) or asymmetrical (sub- and superordination). The social situation in which the partners interact can be private or public. Finally, the speech register can be normative (directive) for a certain situation or partner relation, whereas in others room is left for the free discussion of problems (discursive). The model dialogs incorporated the following combination of pragmatic features: table 2.

The three model dialogs were constructed with comparable

Table 2. Distribution of pragmatic features in model dialogs

Dialog	Hierarchical Features Symmetrical	Asymmetrical	Social Situation Private	Public	Speech Register Discursive	Directive
D1	+		+		+	
D2		+		+	+	
D3		+		+		+

structural characteristics. The macrostructure consisted of an opening stage with a greeting formula and an indication of the direction the dialog would take. The middle part contained the actual negotiation kernel with conflicting points of view and the attainment of a common communicative goal. The closing stage brought the dialog to an end with a verbal stereotype. At the microstructural level, role size and speaker turns (7 to 9) were controlled. The syntactic structure was limited to simple sentences, with approximately 25 % questions per role. In the lexical-semantic material, 40 % of the verbs were without, the remaining 60 % with modal verbs. The dialogs thus constructed were literally played from the script and videotaped.

3.2. Simulation of Model Dialogs

After having watched the videotaped model dialogs, a linguistically healthy partner and an aphasic patient played the dialogs with role change. Following an interval of two weeks, two non-aphasic patients simulated the same model dialogs, also with role change.

The results of this study are based on an analysis of the following dialogs. (table 3).

Table 3. Model and simulated dialogs with partner constellations and number of roles per role type.

	Partner Constellations		
Model Dialog	Model with Model	Non-aphasic with Non-aphasic	Aphasic with Non-aphasic (and vice versa)
D_1	MO_1 with MO_2	NA_1 with NA_2	A_1 with NA_1
		NA_2 with NA_1	NA_1 with A_1
D_2	MO_1 with MO_2		A_1 with NA_1
			NA_1 with A_1
D_3	MO_1 with MO_2	NA_1 with NA_2	A_2 with NA_2
		NA_2 with NA_1	NA_2 with A_2
N° Dialogs	MO with MO=3	NA with NA=4	A with NA=6 (and vice versa)
N° Roles/ Role Type	MO=6	NA/NA=8	A/NA=6

The simulated situation automatically introduced pragmatic features of which some were quite different from those in spontaneous dialogs, others changed the pragmatic characteristics of the model dialogs.

3.2.1. Pragmatic features of simulated and model dialogs different from spontaneous dialogs. Even though the simulated and model dialogs reproduced a realistically possible situation, they were nevertheless acted, which made them answer fixed conditions and diminished the degree of reality. D1 especially differs from the spontaneous situation in this respect. Whereas it would spontaneously happen in private, the subjects in this study met for the first time. The degree of preparation of the partners was also basically different from spontaneous dialogs. Before the subjects watched the model tape, they were told that they would have to play both roles freely later on. After they had been assigned their roles, they were shown the videotape a second time. In order to prepare the role change, they were shown the film a third time. In the model dialogs, the preparation was even more obvious, since they were literally played from a script. In contrast to spontaneous dialogs, the degree of freedom of the topic was fixed. In the simulated dialogs, the subjects has to stick to the course of events and of argumentation of the model dialogs. Word choice was free, however, which was not the case for the model dialogs.
The simulated situation also imported other hierarchical features. Although there was a variation between symmetrical and asymmetrical partner relations in the content structure, the four subjects had a symmetrical relation as fas as their status as patients was concerned. With respect to their communicative abilities, there was a basic asymmetry between linguistically healthy patients and aphasic ones, and a symmetry between the two aphasics.

3.2.2. Pragmatic features common to simulated dialogs, model dialogs, and spontaneous dialogs. The nature of the interaction was face to face in all cases, the partner constellation was diadic and personal, and there was an empractic relation between verbal communication and non-verbal praxis. The latter was always auxiliary and never substantial to the dialogs.

3.3. Transcription

All videotaped simulated and model dialogs were transcribed with a simplified version of the system used by Ehlich and Rehbein (1976). It contained all interactive features necessary for this study: verbal utterances, main prosodic features and gestures, and it excluded irrelevant paralinguistic features such as pitch and intensity. Only those prosodic features with a clear linguistic function were included (e.g. question intonation). Gestures were only noted very selectively also. They were always transcribed when substituting speech but speech accompanying gestures were only integrated if they were essential for the action. Word finding difficulties were defined as pauses of more than 5 seconds, and dysarthric pronunciation was marked explicitly. Sentence interruptions were indicated by dots. Before the partner contributions could be evaluated for interactive characteristics, they had to be divided into the relevant utterances, i.e. the units of analysis had to be defined. The exclusive use of propositional-semantic criteria led to the establishment of units which were too varied in their syntactic form: they could be interjections, simple or complex sentences, or even several sentences which were semantic variations of each other. The additional use of syntactic criteria could set an upper limit which was also semantically valid. The clauses in a paratactic construction are semantically more independent of each other than those in a hypotactic one. Therefore, coordinating clauses were considered as separate units, whereas the entire complex sentence in the case of subordination was considered to be a single unit. Fragmentary sentences were treated in analogy to the nearest reconstruction. Particles, which are very frequent in German, were evaluated by means of semantic pragmatic criteria. They were considered a unit by themselves if they constituted a turn or if, within a turn, they had an independent semantic function. The units of analysis thus defined were integrated into the transcription line, and the units were consecutively numbered per speaker. If the two partners spoke simultaneously, their lines would be filled in parallel. If turn taking was maintained, the transcription would show this by having one partner's line start where the other partner's transcription line ended.

4. DATA ANALYSIS

The transcribed units were evaluated for their interactive dialog function by three examiners and classified into one of the following main categories:

- content units, relating exclusively to the dialog. Subunits included here are dialog opening, closing, and discontinuation; introduction, variation, and continuation of a topic; asking the partner's opinion and giving evaluative feedback.
- Conflict units, relating to the negotiating character of the dialogs. Subunits included here are conflict build-up and proposal of possible solutions, and the repetition of one's point of view.
- Language units, relating to the linguistic form of the partner's utterance. Subunits included are corrective feedback of an ambiguous partner utterance, feedback with sentence or sequence completion, anticipation of a possible partner reaction with a simple sentence model which could be used as a response by the partner, repetition of the partner's utterance, and redundancy.

- Mixed units, relating to both dialog content and linguistic form. Subunits included are questioning of the partner's intention and anticipation of the partner's reaction with a sentence model too complex to be used as a response.
- Unclassifiable units, the nature of which could not be agreed upon by the three examiners.

The three model dialogs and ten simulated dialogs (4 between) non-aphasics, 6 between an aphasic and a non-aphasic) were analyzed in this way. There were thus 8 non-aphasic roles in dialog with each other, 6 aphasic roles in dialog with non-aphasics, and 6 non-aphasic roles in dialog with aphasics. The units were distributed over these roles as follows: see table 4.

RESULTS AND DISCUSSION

On the basis of informal observations, the following results could have been expected:

1. The non-fluent Broca's aphasic patient deviates not only in his linguistic performance but also in his dialog behavior. This is caused by and secondary to his impaired expressive language, which prevents him from asserting himself verbally. As a consequence, his strategies of verbal negotiation will be impaired.

This can be clearly observed in dialogs of non-fluent aphasics with each other, where the partners are not able to react to each other's dialog contributions and argue their own point of view convincingly. The dialog will come to a premature end often when conflict is encountered and always before a solution has been found. The disturbed verbal interaction pattern of the aphasic patient can also be observed in dialogs with normal partners who are either uncooperative or overprotective - as relatives and speech therapists typically tend to be. They will tend to domineer

Table 4. Distribution of classifiable units over 4 dialog categories per role type.

Number and Percentage of classifiable units per dialog category per role type

Category	Language		Conflict		Content		Mixed		Total Classifiable	
Role Type	No	% Total	No	% Total	No	% Total	No	% Total	No	% Total
MO	2	2	23	23.5	71	72.5	2	2	98	95
NA/NA	15	6	70	25.5	181	68	2	0.5	272	94.5
NA/A	25	11	42	19	142	64	12	6	221	89
A/NA	9	8	19	16	86	73.5	3	2.5	117	75

the patient verbally, who will finally be pushed into retreating into at best an affirmative corner. The course of the dialog is then severely disturbed, with length and number of turns being drastically reduced in the impaired partner.

2. The communicative strategies of the non-aphasic partner in dialog with an aphasic partner will be different from that of non-aphasics with each other. Their hearer role is impaired as a consequence of the aphasic partner's expressive disorders. If he is cooperative, he will try to safeguard the dialog rules by using the necessary repair mechanisms.

In an observation of a simulated negotiation dialog between a cooperative non-aphasic and a severely non-fluent aphasic, it has been shown that the non-aphasic used specific linguistic repair mechanisms, such as redundancy, anticipation of the partner's reaction, corrective feedbach of an ambiguous partners utterance, etc. (De Bleser-Weismann, 1981). In addition to such direct feedback strategies, more indirect mechanisms such as questioning the partner's intention were also used.

On the basis of the data in table 4, a test of homogeneity of the four role types was made with respect to the four categories (Gabriel, 1966). The distribution was not homogeneous ($\chi^2 = 26.9 > \chi^2$ (9.95%) = 16.9). Pairwise comparisons were then made of the role types. These showed a significant difference between the non-aphasics in dialog with the aphasics (NA/A) from dialogs with other non-aphasics (NA/NA). All other comparisons were not significant, including those of the aphasics (A/NA) with the model dialogs (MO) and with normals (NA/NA).

Pairwise comparisons of the categories with respect to the two significantly different role types (NA/A versus NA/NA) demonstrated a clear - though not yet significant - difference for mixed versus content categories. In comparison to non-aphasics

Table 5. Results of pairwise comparisons of the four role types

Role Type	MO	NA/NA	NA/A
NA/NA	3.5		
NA/A	12.1	17.9*	
A/NA	5.1	6.3	3.8

* χ^2 (9.95%) = 16.9

Table 6. Results of pairwise comparisons of the four categories for role types NA/NA versus NA/A

Category	Language	Conflict	Content
Conflict	7.4		
Content	5.2	1.2	
Mixed	2.8	12.4	10.3

χ^2 (9.90%)= 14.7 χ^2 (9.95%)= 16.9

with each other, a non-aphasic in dialog with an aphasic partner uses more mixed categories and less conflict categories. Descriptively, the increased use in mixed categories can be retraced to the category of "questioning the partner's intention". This was the only category filled by units in the dialogs examined. This category reflects the partner's attempt to check whether he understood the other's contribution correctly concerning both content and form. It shows the indirect repair the non-aphasic makes especially by means of intonation questions which are simultaneously yes/no questions. With a minimal reaction of the patient (affirmation or negation), the dialog can thus be continued. Evidence for an increased use of more direct repair mechanisms, as represented by the language categories, could not be found. This is partly due to the use of two sub-units, redundancy and repetition of the partner's utterance, in dialogs of non-aphasics with each other as well (NA/NA). However, these were the language categories used exclusively in this constellation, whereas in dialogs with aphasics (NA/A), the remaining language categories were also filled by units.

The limited use of these direct repair mechanisms may reflect the tendency of the non-aphasic to intrude as little as possible in the course of the dialog. In the informal observation made earlier (De Bleser and Weismann, 1981), the patient was much more

Table 7. Distribution of language subcategories in the role types NA/NA and NA/A

Language subcategories	Role Type NA/NA N°	NA/NA %total	NA/A N°	NA/A %total
Redundancy	11	4	14	6
Repetition of partner's utterance	4	1.5	4	2
Others	--	--	7	3

impaired expressively than the Broca patients in this study, who were not dysarthric and had far less speech and language effort than the previous patient. Given this relatively better language expression, the unimpaired speaker could guess the intended message and could verify whether he should continue his contribution with the partner intention he had interpreted, or whether he should first correct it.

The cooperative attitude of the non-aphasic was not only shown in the use of an indirect repair category, but also in his employment of other dialog features. Although his linguistic superiority could easily allow him to take over the lead and monopolize the floor, he does not exploit this possibility. The amount of sequences per dialog, the number of turns and of units produced per speaker in the dialogs varies as much for non-aphasics with each other as with an aphasic patient. A finer linguistic analysis would additionally reveal that the unimpaired partner also adjusts non-intrusively by using shorter and simplified sentence structures, which are easy for the aphasic to understand and to use as a model for his production. Such delicate communicative and linguistic adaption of the unimpaired cooperative partner makes it possible for the aphasic patient to contribute actively to a negotiation dialog in a pefectly normal way from an interactional point of view, even though his means of linguistic expression are quire disturbed. It is only when the expressive means of the aphasic partner are insufficient that the normal partner will resort to more direct repairs and to longer speaker turns. In this case, the aphasic patient not only disturbs the hearer role of the normal partner but actually makes it impossible, so that interaction would come to an abrupt end without direct intervention.

The disturbance of verbal interaction which normally follows from the impaired language system of the Broca's aphasic will, therefore, only demonstrate itself clearly when the partner is uncooperative, intentionally as may be expected in real life, or unintentionally as with another aphasic. If the non-aphasic partner agrees to the dialog objective and is interested in a balanced interaction, as was the case here, he will spontaneously reduce his linguistic and communicative superiority. He tries to accomplish this, not by speaking foreigner or baby talk, which would be condenscending to the dialog partner, but by normal means of linguistic simplifications and by non-invasive, indirect means of dialog repair.

REFERENCES

Boller, F., Green, E., 1972, Comprehension in severe aphasics. Cortex 8, 382-292.

Brownell, H., Michel, D., Powelson, J., Gardner, H., 1983, Surprise and coherence: sensitivity to verbal humor in right hemisphere patients. Brain and Language 17.

De Bleser, R., Weismann, H., 1981, Übergang von Strukturübungen zum spontanen Dialog in der Therapie von Aphasikern mit nicht-flüssiger Sprachproduktion. Sprache, Stimme, Gehör 5, 74-79.

Ehlich, K., Rehbein, J., 1976, Halbinterpretative Arbeitstranskription. Linguistische Berichte 45, 21-41.

Gabriel, K., 1966, Simultaneous test procedures for multiple comparisons on categorial data. Journal of the American Statistical Association 61, 1081-1096.

Goodenough-Trepanier, Ch., Powelson, J., Zurif, E., 1982, Bridging in right hemisphere patients. Paper presented at the 20th Academy of Aphasia, Lake Mohonk, N.Y.

Henne, H., Rehbock, H., 1979, Einführung in die Gesprächsanalyse. De Gruyter, Berlin-New York

Poeck, K., 1978, Neurologie. 5th Edition. Springer, Berlin-Heidelberg-New York.

Ulatowska, H., Doyel A., Freedman Stern R., Macaluso Haynes, S., North, A., 1983, Production of procedural discourse in aphasia. Brain and Language 17.

Wapner, W., Hamby, S., Gardner, H., 1981, The role of the right hemisphere in the apprehension of complex linguistic materials. Brain and Language 14, 15-31.

Zurif, E., 1980, Language mechanisms: a neuropsychological perspective. American Scientist, 68, 105-111.

DYNAMICS OF INTERACTION IN SPEAKING DIALOGS WITH DEAF CHILDREN IN THE CLASSROOM

Filip Loncke and Marijke Van Weerst

Lokeren

In the literature on deafness, much attention is paid to the question of how language is acquired. It is clear that this is not simply a matter of having a good balanced didactic methodology. It is important to have a strong theoretical view on the way language acquisition develops and can be influenced. From such a view implications can be deduced about the style teachers should adopt in their communication and didactic settings with deaf students.

Already in very early studies the suggestion is made that teaching the deaf should be inspired by the way mothers communicate with their young hearing children. In 1860 Dalgarno pointed out that deaf children should learn their language in the same way as hearing children acquire their mother-tongue, and to facilitate this the mother should fingerspell to the deaf baby from birth. (In Savage et al., 1981, p. 9)

With this statement, Delgarno happens to be a precursor of a strong educational movement in favour of the so-called natural language acquisition teaching method for young deaf children.

While other methods that have been developed to teach or train deaf children to use the spoken language - such as the imitative and the constructive methods (vor a review see e.g. Van Uden, 1977) the strength of the natural methods obviously lie in the fact that they seek their motivation in observations of the development of communication and spoken language.

The method therefore has to be evaluated against that background. The main questions that arise concern the exact interpretation of the data available through language acquisition studies and whether of the extrapolation to deaf children and to didactic situations can be made.

However, in many respects school seems to be drastically different from the climate on the family where the very young child acquires the basic rules of syntax, vocabulary use and semantics. This difference obviously lies in pedagogical style and in implicit and explicit educational goals. It can be interpreted as a difference in terms of pragmatics. In this paper we will focus on the validity of the ideas behind the possibility of extrapolating the primary communicative linguistic situation to didactics. On the basis of a pilot study we undertooks in a class for deaf children and in one for hearing children, we will try to formulate precise predictions on the limits set by a so-called natural language acquisition teaching method for the deaf.

We aim to do this by investigating some parameters which we consider to be central in language stimulation: age, didactic style and pedagogical goals and prior linguistic knowledge of the subjects.

1. LINGUISTIC KNOWLEDGE OF THE SUBJECT

It is of course very important to have a clear view of the state of the art concerning the linguistic knowledge of deaf pupils at any moment of educational intervention.

In most cases in Western countries considerable efforts are undertaken to detect deafness as early as possible. It is recognized that the earlier the interventions can start the higher the chance for a succesful education.

Most of the time this success is partially or entirely measured by the degree of mastery of the cultural "spoken" language.

Of course, it is very important that deaf children master spoken language in its spoken and written form. However, by stressing this central point, there is a tendency to overlook on the one hand the roots of communicative and linguistic behaviour in both deaf and hearing children and, on the other hand, the strong linguistic capacity of children which, in the case of the deaf, tends to be actualized primarily in sign form.

Modern developmental psycholinguistics no longer sees the early vocalisations and vocal playings of the baby as the sole precursors and preparers of language. A very important part seems to be played by social interactive processes which form a basis for the symbolic-referential, the communicative and the pragmatic aspects of language (Halliday, 1975).

The early interaction between mother and child in daily ritual situations such as feeding and washing contain the ingredients of symbol formation, referential use of it and turn-taking (Bates, 1976). The linguistic symbol and its use is built on similar interactive early situations in both deaf and hearing children. An essential characteristic of the symbol however is its multimodal nature: it can be a visual-gestural, a mimic, a vocal, a spoken symbol or a combination of two or more of these modes.

Due to social development the preference for the spoken mode in hearing children becomes more and more clear after his first year (Volterra, 1979).

In deaf children, although they have a similar experience in communication during their first year, the selection of the spoken symbol as the predominant one is less evident.

On the contrary there is growing evidence that deaf children develop a visual-gestural code which very quickly surpasses the limitations of early context-bound communication by displaying a strong combinatorial capacity in the visual-gestural mode (Volterra, 1981) or - when the children have no adult models who systematically use signs - by using the gestural symbols in linguistic ways that surpass the structures the adult model shows (Goldin-Meadow, 1982).

When the children do have deaf parents there is a fair chance that they will use some form of sign language. It has been shown that this often turns out to be a favourable condition for linguistic, social and general psychological development.

In the last decade deaf children of deaf parents have been investigated from the point of view their psycholinguistic development in sign language. This research has proven to be very surprising in that the development of linguistic structures in the visual-gestural mode parallels the spoken language acquisition in young hearing children, but it also offers growing evidence for a slight advance in the appearance of the first linguistic symbol in the gestural mode (Bonvillian, 1983).

If the underlying hypothesis of the teaching method for deaf children called the "maternal method" is that they have to acquire the spoken language as their first language code, it seems to be refuted in the light of this evidence.

It remains to be checked if an educational policy which considers that children with a relative linguistic fluency in the visual mode must still have to acquire true linguistic skills can be successful. Anyway, it does not take into account the possibility of making use of the linguistic strength of the gestural symbol.

2. AGE-RELATED FACTORS IN LANGUAGE STIMULATION

Although the overage moment of discovery of deafness seems to happen much earlier than a few years ago (Layon-Verstraeten, 1982) due to a better detection system, enabling an earlier treatment of deaf children together with guidance of the parents, the delay in the production of the first spoken words and sentences between deaf and hearing children is nonetheless still striking.

Different authors give different figures on the spoken vocabulary and the syntactic skills of very young deaf children. It is clear that these differences are at least partially due to the adequacy of the pedagogical method.

However, it is apparent that the gap between the first appearance of linguistic skills in the manual mode and the first spoken communicative performances is quite impressive. A problem arises when one tries to seek inspiration for spoken language stimulation in the communicative situation between the very young hearing child and his mother. It has been demonstrated (Snow & Ferguson, 1977) that this communicative situation has very specific characteristics, which are probably generated by the very nature of the communicative situation itself. A central concept in this matter is the existence of Baby Talk, which turns out to be a very refined and sophisticated tuning-in in relation to both the cognitive and linguistic abilities of the child. Part of the setting of the first mother-child situation in which Baby Talk is an interesting phenomenon, is the age of the child.

Any extrapolation of the characteristics of this early mother-child communication should therefore take into account this age-related cognitive factor. In other words, the suggestion that the teacher should use the mother-child sitiation as containing the guiding principles for the stimulation of the deaf child's spoken language risks to overlook not only that the child has already been very active in displaying linguistic performance in the visual-gestural mode, but also the very fact that their cognitive evolution has long surpassed the level of the hearing child in whom the mother is stimulating the first spoken utterances.

Transposing the idea of maternal stimulation of spoken language deaf children at primary school level run the risk of creating an inadequate means because it forces the teacher to use a methodology which is part of a phenomenologically (and psychologically) totally different situation. In the following we will elaborate this point.

3. DIDACTIC PURPOSES AND DIDACTIC STYLES

Contrary to his colleague working with hearing children the teacher of deaf children is facing a double task: on the one hand he has to initiate his pupils in the basic elements of the culture, including reading, writing and arithmetic skills, and on the other hand he has to stimulate them to acquire adequate attitudes enabling a socially adjusted and self-reliable behaviour in society. The teacher of hearing children as a rule doesn't have to worry about the basic acquisition of a communicative code to convey information.

The problem with the teacher's task as language stimulator obviously lies in the fact that he has a double role: he has to be concerned about the linguistic forms the children produce and at the same time he must convey cultural content.

By focusing on the form and giving feedback, one can speculate that he may overlook the child's need to have information about the world. Because of his deafness the child is already less informed about his surroundings. If one then adopts a philosophy of language stimulation which consists of giving systematically a feedback on the linguistic form the children produce, one could wonder if an extra slowdown in the didactic process may not occur. This appears to be a paradox in the education of the deaf. One possible solution is to expect that the linguistics-centred pedagogy will be fruitful after some time in yielding two important effects: on the one hand the child may have acquired a good mastery of the spoken language which is clearly an encouraging perspective and on the other hand by this method a solid way is paved for conveying information. The weakness of this reasoning lies in the fact that it takes it for granted that the communicative style between a very young hearing child and his mother or mother-figure (caretaker) has some specifically identifiable aspects that intrinsically stimulate language expansion has been mentioned and has been suggested to be extremely valuable for linguistic development. In handbooks on deaf education it is therefore suggested that the teacher of the deaf should systematically make use of expansion.

However, its function is less certain than is often supposed:

"... Recent language-acquisition proposals hypothesize that the language environment of the child becomes succesively more complex in correspondence with the child's growing language skills, and thus may be at all times appropriate for an environmentally-dependent acquisition process. For example, Levelt (1975) writes 'the child is presented with grammatical strings from a miniature language, which is systematically expanded as the child's competence grows'. Our findings suggest instead that many features of the mother's speech change in accordance with the child's age, not his competence with constructional features of the language". (Newport et al., 1977, p. 145).

This hypothesis seems to us to have possible important consequences on deaf education practice. If the typical aspects of communication with very young children who are not yet in command of the spoken language code, lies not only in the fact that they are still linguistically naive, but also in that they are at the threshold of cognitive exploration, then the extrapolation toward a recommendation of a certain didactic style loses its validity.

4. INTERACTION PATTERNS IN DIDACTICS WITH DEAF CHILDREN

In the former paragraphs we have formulated some criticisms about psycholinguistic motivations concerning teacher style in communication with deaf pupils.

On the one hand, the transposition of what happens between the very young hearing language acquirer and his mother toward the spoken language stimulation of deaf children in school is problematic for both cognitive and linguistic reasons.

The deaf child at primary school has already reached a level of concrete operational thinking (Furth, 1966, Bartin, 1981). This is of course a much higher level than the hearing child when he starts to talk. The deaf school child has also a relative mastery of linguistic skills, mainly in the manual mode. Although his spoken fluency is mostly weak, he has a certain linguistic awareness and he uses linguistic deictic and referential (manual) symbols in combination (Volterra, 1981).

On the other hand in the classroom an atmosphere is supposed to stimulate linguistic, cognitive and social development.

One can formulate different hypotheses about the way teachers and pupils cope with such a situation. One hypothesis is that the different goals are compatible and that any didactic action of the teacher at any moment can be interpreted as favouring all three goals.

A second hypothesis is that the different goals cannot be served at once and that there is a flexible interchanging of prevalence of the different goals.

A third hypothesis is that one or two aims are so prevalent that the other one or two are hardly touched on.

In the rest of this paper we will discuss these hypotheses on the basis of pilot study of some didactical processes used in classes for deaf and for hearing children. Due to the limited amount of data we do not intend to make definite conclusions. Instead we hope to find evidence to prove the validity of extrapolating psycholinguistic language acquisition phenomena of the young hearing child to the formulation of a phenomenology for teaching deaf children.

5. TEACHING THE DEAF

What really goes on in classrooms for the deaf? Most teachers of the deaf in Western Europe have been trained in techniques which foster speech and stimulate the acquisition of the spoken language code. These consist of eliciting spoken conversations of daily experiences, giving reinforcement to attempts at spoken utterances, the use and the modulation of the voice, the systematic use of expansion, exercising in reflections on the form and the structure of the spoken language and the early introduction of reading ("ideovisual reading") (see e.g. Van Uden, 1977, Moores, 1982).

In our experiment, we wanted to check the validity of the maternal method in the teaching of the deaf. In the following section we will use the term "maternal hypothesis" for the supposed ability of the teacher (1) to reinforce the deaf child's verbal utterances by expansions, while (2) guiding the didactic process.

This could be translated into a kind of pragmatic process, in which certain behavior patterns in the teacher will probably be prevalent, such as:
- "grooping" what the child is trying to say.
- reformulation in "expanded form" of what the child tried to say formerly
- stimulation of utterances in the child, e.g. by asking questions (high percentage of pupil talk)
- making contributions to a conversation which will ensure its continuation.

To check if these behavior patterns really appeared, we set up a comparison of didactic styles between a teacher of the deaf and a teacher of hearing children. The teacher of the deaf had ten years experience and had been trained in the methodic style of interacting with young deaf children using primarily speech and written language.

The deaf class consisted of eight children (4 girls and 4 boys), most with a fair ability in the mastery of spoken language. This was a socalled "oral group", although sign communication was accepted and used as a valuable tool, mostly for non-didactic situations. All children had normal intelligence scores (SON IQs higher than 95) and no additional handicap.

The group of hearing children varied from 10 to 14 (4 girls and 6 - 10 boys). All children had normal intelligence. The ages varied from 11 to 12 years.

The teachers agreed on teaching lessons each 20 minutes long on the same topics, generally based of deduction and classroom conversation. The topics were:

1. geography: the lesson aimed at gaining knowledge of the names of the different parts of a river (exactly the same visual didactic material was used).
2. physics: the notion of "air-pressure" was explained and demonstrated using a series of small tests in physics.
3. vocabulary: words unknown to the children were explained and taught to both groups. In the group of deaf children the topic was "springtime" and in the hearing it was "carnival".
4. mathematical questions on percent computation: the groups were confronted with three problems of which two were the same for both groups.

Twenty minutes of each lesson was analyzed using an interaction scheme based on Flanders' (1970) interaction matrix, but extended with categories relevant to our problem (see appendices A and B). The "maternal hypothesis" predicts a high frequency of specific behaviour in the teacher of the deaf children, compared with his colleague in a class of the hearing (grasping-meaning, expanding, ...).

Each second the interactive process was coded into one of the proposed categories. Based on this observation, we were able to construct flow-charts of the lessons.

5. Results:

It appears that the interaction in the deaf class is less varied than in the hearing class. In the diagram less cells are occupied than in the hearing group's diagram.

In the deaf class the percentage of teacher Talk is higher than in the hearing group. It seems that the teacher of the deaf has the feeling that he has to keep the initiative permanently in the classroom. It is also noteworthy that certain categories are more prevalent in the deaf class: more praising and more self-repetition by the teacher. This may be conceived as language stimulating or communication-ensuring mechanisms.

Interesting data are found in the pupil talk. In comparison with their hearing counterparts, deaf pupils have less pupil talk. This is even more surprising since the teacher of the deaf asks more questions. However, stated in terms of didactic efficiency, deaf children are much more accurate in their answers.

This suggests another aspect of the deaf condition and its

implications for didactics: a tendency toward verbal shortness and accuracy and less verbal (spoken) fluency.

An analysis of the pragmatic situation with a cummunication flow-chart study shows that the teacher of the deaf happens to be much more directive in the didactical process. He often repeats the answers of the children literally.

However, the teacher of the deaf shows a somewhat different behaviour in his vocabulary- lesson. While in his other lessons he seemed to be preoccupied by playing his role as leader of the didactic process, he now leaves this strategy and evokes more answers of his pupils.

6. DISCUSSION

Textbooks on deaf education tend to emphasize the teacher's role as language stimulator. While the spoken language of deaf school children may be reminiscent of the spoken language level of very young hearing children, it may be misleading to take this resemblance as the starting point for a deaf educational didactic program, both for linguistic and cognitive reasons.

The maternal approach seems to fail to take into account the very nature of the classroom situation. In our pilot study it appeared that this situation entails a whole set of interactive behavior patterns (a.o. directed towards didactic effectiveness and characterized by a accuracy- cognitive style) which are phenomenologically very different from the mother-child situation. Only in language lessons was the pattern a little more "maternal-like".

7. REFERENCES

Bartin, M., 1981, Etude Comparative de l'Atteinte du niveau opératoire formel par des subjets sourds profonds de naissance, Enfance, 386-390.

Bates, E., 1976, Language and Context. The Acquisition of Pragmatics, New York, Academic Press.

Bonvillian, J., 1983, Language, Cognitive and Cherological Development: The First Steps in Sign Language Acquisition, Paper presented at the IIId International Symposium on Sign Language Research, Rome, 22 - 26 june.

Flanders, N., 1970, Analyzing Teaching Behavior, Reading, Mass., Addison-Wesley Publishing Company.

Furth, H., 1966, Thinking without Language. Psychological Implications of Deafness, New York, The Free Press.

Goldin-Meadow, S., 1982, The Resilience of Recursion: a Study of a Communication System developed without a Conventional Language

Model, in Language Acquisition: the State of the Art, Gleitman and Wanner, eds., New York, Cambridge University Press.
Halliday, M., 1975, Learning how to Mean, London, Edward Arnold.
Layon-Verstraete, E., 1982, Opvang en Begeleiding van Auditief Gehandicapte Peuters en hun Ouders, Tijdschrift voor Logopedie en Audiologie, 12, 3, 174 - 190.
Moores, D., 1982, Educating the Deaf. Psychology, Principles and Practices, 2nd edition, Boston, Houghton Mifflin Company.
Newport, E., Gleitman, H., Gleitman L., 1977, Mother, I'd rather do it myself: some Effects and Non-Effects of Maternal Speech Style, in Talking to Children, Snow & Ferguson, eds., Cambridge Cambridge University Press, 109-149.
Savage, R., Evans, L., Savage, J., 1981, Psychology and Communication in Deaf Children, Sydney, Grune and Stratton.
Snow, C., Ferguson, C., 1977, Talking to Children. Language Input and Acquisition, Cambridge, Cambridge University Press
Van Uden, A., 1977, A world of Language for deaf Children. Part I: Basic Principles. A Maternal Reflective Method, Lisse, Swets and Zeitlinger.
Van Weerst, M., 1983, Interactie tussen Dove Lagere Schoolkinderen en hun Leerkracht. Een vergelijkende Gevalstudie, Dissertation VUB.
Volterra, V., 1979, Symbolic Development in Spoken and Gestural Modalities, Paper delivered at the Nato Congress, Copenhagen
Volterra, V., 1981, Gestures, Signs and Words at two Years: When does Communication Become Language? Sign Language Studies 33, 351-362.

APPENDIX A: THE CODING CATEGORIES

TEACHER CATEGORIES

Category no	Description	Responsive (R) or non-reps. (nR)	Criterion
1.1.	Interprets or accepts feelings	nR	4
2.1.	Praises	R	4
2.2.	Praises, but demands additional effort	R	3 and 4
3.1	repeats literally	R	2

3.2.	repeats the pupil in a re-formulated way	R	2
3.3.	verbalizes an action	nR	2
4.1.	asks a question oriented toward the subject-matter	nR	4
4.2.	asks check-question	R	4
4.3.	asks "fishing" question	nR	4
5.1.	lectures	nR	-
5.2.	answers himself	nR	-
6.1.	gives a task	nR	-
6.2.	gives additional information	R	3 and 4
7.1.	criticizes	R	3 and 4
7.2.	self-criticizes	nR	-
11.1.	literally self-repetition	nR	4
11.2	self-repetition in a reformulated way	nR	4
12.1.	starts new topic	nR	-

PUPIL CATEGORIES

Category no	Description	Spontaneous or Non-Spontaneous
8.1.	Answers in a predictable way	NS
8.2.	repeats what the teacher said	S

(Cont'd)

TEACHER CATEGORIES

8.3.	anticipates	S
9.1.	introduces a new idea	S
9.2.	asks for additional information OR asks for an intervention	S
13.1.	reinforces the teacher	S

OTHER CATEGORIES

Category no	Description
10.1.	Silence
10.2.	Confusion

APPENDIX B: THE MATRIX

	1.1	2.1	2.2	3.1	3.2	3.3	4.1	4.2	4.3	5.1	5.2	6.1	6.2	7.1	7.2	11.1	11.2	12.1	8.1	8.2	8.3	9.1	9.2	13.1	10.1	10.2	
1.1	1				1					1															1		Σ 4
2.1		1	1	2	5		5	1	2	8		2					2	1	3		2	1			5		Σ 41
2.2			2				1			1									5				1				Σ 10
3.1		8	4	8	2		2	2	2	7				8		5	1		8	5				2	7		Σ 71
3.2	1	8	1		68		9		2	10		1				2	4		2		3	6	2		7		Σ 126
3.3							1			1																	Σ 2
4.1						1	50	1		2		6				7	6		20				14		8		Σ 115
4.2								6			1			1		1			13			3	3		2		Σ 30
4.3							1		6			3							4				3		5		Σ 22
5.1		2		1			8		1	169	1					6	9	1	5	10	7	8	10	7	22		Σ 267
5.2								1			1					2				1					2		Σ 7
6.1		1						1		1		2							51	2	1	1	1		5		Σ 66
6.2																			1								Σ 1
7.1					1	1	2		1	3		1		14		2	2		5				2				Σ 34
7.2																											Σ 0
11.1				1	3		3		1	6						22	4		5	5	1	3	5	1	6		Σ 66
11.2	1	1					6		1	7		1				4	60		2	4	3	3	2		11		Σ 106
12.1							1											6							1		Σ 9
8.1		10	2	43	27		2	11	1	6	3	15		10		2	1		112	1			2		11		Σ 259
8.2		5		7	6		2	1		10		2				1	2			22	2	1	1		3		Σ 65
8.3				5	3		3		3	2		1				2					12		1		1		Σ 33
9.1	1	3		2	4		2	3	1	5		1				1	1					39	3		4		Σ 70
9.2					1		7			5		24	1				2		7			2	4		6		Σ 59
13.1				1				1		3						1	1		1	1	1						Σ 10
10.1		2		1	5		10	2	1	19	1	7		1		8	11	1	15	14	1	3	5		75		Σ 182
10.2																											Σ 0

EVOLUTION OF AN APHASIC CHILD AFTER THE INTRODUCTION OF NVCDs

F. Lowenthal

Laboratoire NVCD
University of Mons

J. Saerens

Department of Neurology
A.Z. Middelheim

1. INTRODUCTION

Acquired aphasia is the loss of language as a result of a brain lesion. The prognosis for recovery from childhood aphasia is variable and is determined by at least four factors: etiology, location of lesion, age at time of lesion, and associated disturbances (Woods, 1982). The goal of therapy for aphasic children is different from that in adults: not only must former language abilities be recovered, but further language acquisition must also be continued.

Several kinds of language therapy have been proposed; most of them are strictly verbal in nature. By making use of special techniques, articulation and grammar are trained in an explicit and drill-like manner. Other aphasic children have been trained to use a symbolic language like sign language or the BLISS-symbols in order to improve their communication abilities.

However, some children with acquired aphasia who do not respond to any of these therapies. The fact that these children do not learn a language, nor any of the alternative communication systems makes it very difficult to differentiate between acquired aphasia and acquired childhood dementia. One of the questions involved here is whether cognitive deficiencies explain linguistic disabilities or whether linguistic deficiencies prevent the child from showing his cognitive possibilities in an appropriate way (Siegel, 1982).

Furthermore, even for the children who fully recover, the

use of verbal and analytic strategies in certain kinds of problem solving exercises remains impaired; this could explain the failure of these children in academic achievement (Denckla, 1979; Van Dongen, 1976; Van Hout, 1983). Childhood aphasia is therefore not only a medical and speech therapeutical problem, but also a pedagogical one.

RENARD (1985) points out that there are prerequisites to the autonomous learning of a language. Some of these prerequisites are of a logical nature: sorting activities first, followed by the discovery of relations between elements of the "universe" of discourse, and then the construction of structures which organize the set of known "things" into a framework within which the subject can actually live.

Logical prerequisites are needed, but it is obvious that they cannot be introduced via a verbal argument since these logical structures are part of the verbal language which is not yet mastered by the learner: the logical structures are thus, at the start, not yet integrated by the learner. This is why we tried to find out how they could be introduced with very little recourse to verbal language. In doing so we defined a technique which we called the technique of Non-Verbal Communication Devices (NVCD). We used it with normal children and with handicapped children who were unable to communicate in a structured way (Lowenthal, 1982).

Using NVCDs with normal children seems to improve their communicative behaviour; using NVCDs with handicapped children made it possible to create the basis needed for the development of a structured communication.

In this paper we will show how NVCDs have been used with an aphasic child, who was totally unable to communicate in a structured way at the age of 5; we will describe this child's evolution and show the results obtained by using this method.

The paper will be divided into seven parts. Firstly we will give a short definition of the concept 'NVCD': further details can be found in LOWENTHAL's paper (this volume); then the case history will be presented; in a third part we will describe the material and in a fourth part the method we used; the results obtained while using NVCDs will be presented in a fifth part and other results will be presented in a sixth part; finally before we conclude and present our future projects, all these results will be discussed in a seventh part.

2. DEFINITION OF A NON-VERBAL COMMUNICATION DEVICE

"An NVCD is essentially a tool (or a set of tools) which is furnished with technical constraints. These constraints make certain actions possible and others impossible; from these facts a logical structure is suggested". (Lowenthal, 1984b). We have tried to use different kinds of devices and have noticed that some are more useful than others when trying to favour the logico-cognitive and communicative development of children. This has led us to the following definition of an NVCD; "it must be a set of tools which satisfies the following requirements: one should be able to confront the child with logical problems in a nearly non-verbal way; there should be no ambiguity; the device must be simple and easy to handle; the simplicity of the device should not restrict the quantity of different messages which could be formulated: thus the system must be flexible; one should be able to use it in the framework of games and finally the device must be furnished with technical constraints. It is important to notice that in order to use such a system one does not need long training".

It is easy to find a device which can be used as an NVCD; the main thing is to use it properly: one should use an NVCD to confront a child with a logical problem which is "just slightly more difficult" than what he already knows, without being "out of reach". By doing this we challenge the child without ever putting him in a situation where he must notice his own failure. The fact that we avoid verbal instructions as much as possible enables us to avoid conflicts with a still developping verbal language.

3. CASE HISTORY

a. Medical History

Saïd S. is a Moroccan boy born on May 13, 1976 and living in Belgium. He is the fourth child of a family of six children: all the other children are normal. His psychomotor development was normal and the boy started to speak and walk at the age of 1 year.

He suffered from a herpetic meningo-encephalitis at the age of 13 months. As a result, the child suffered for a short time from a right hemiplegia. He recovered, but there was still a slight right apraxia at the age of 5 years 6 months, when we examined him for the first time. This apraxia seems to have nearly, but not completely, disappeared now at the age of 8,6 years for unknown reasons but after the subject started using NVCDs. Saïd still has severe swallowing problems due to the paralysis of some of his cranial nerves (pseudo-bulbar syndrome). He behaves like an intelligent boy, goes alone to the toilet, and recognizes people.

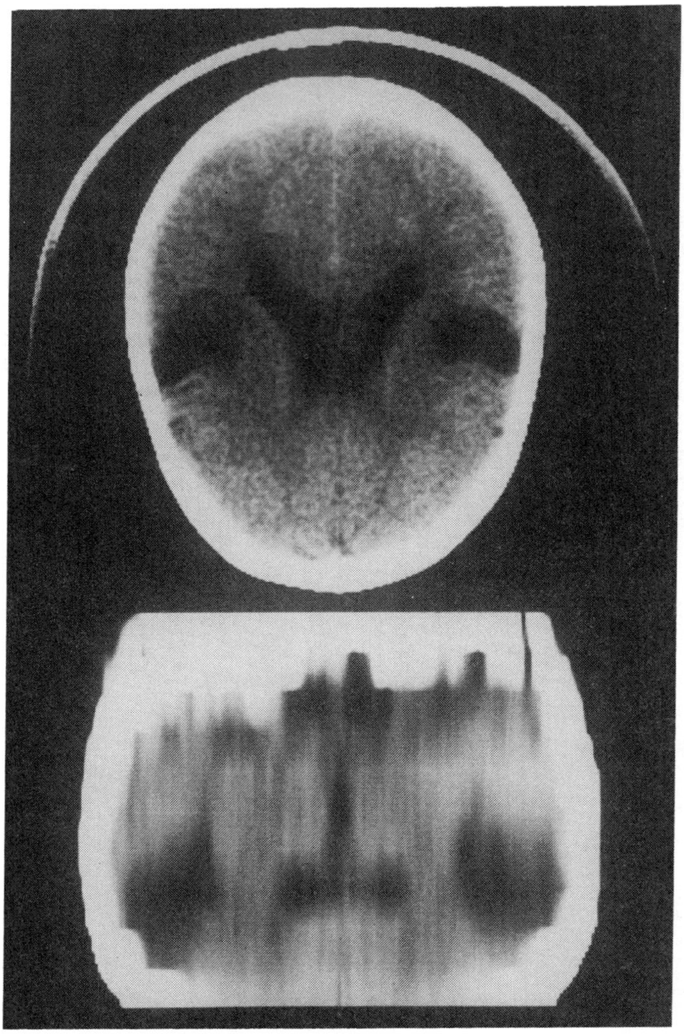

Figure 1. CT-scan of Saïd at the age of 6,5 years.

Formal hearings tests, tonal audiogram and BERA have always been normal. His EEG is normal. Repeated CT-scans show two stable, symmetrically localized lesions in the Rolandic and parietotemporal regions (figure 1).

b. Previous Treatment

Since the age of 13 months, Saïd has been treated by several speech therapists and examined by several psychologists. The child was not able to utter or understand spoken words. Psychologists considered him as severely mentally retarded: the child was not able to answer their verbal questions. The speech therapists tried to teach him to use BLISS-symbols or the signs used by deaf mutes; his apraxia made this difficult: altogether this was a complete failure.

In fact he was unable to imitate complex gestures, so that he could not be taught a gestural communication system despite the fact that the boy tended to make a few indicating gestures spontaneously. He had a kinetic apraxia, i.e. he was unable to start the movements of drawing although he had no paresis of arm and hand. When he was taken by the hand to start the movement, he could continue by himself.

c. First Neurolinguistical Examination

We examined Saïd for the first time at the age of 5,6 years. He was unable to understand and speak a natural language, indicating a global aphasia, complicated by a mild auditory agnosia: the boy found it difficult to differentiate between acoustically similar noises like the ringing of a door-bell and the whistling of a flute. He could not produce voluntary sounds and even laughed and cried in silence. He could not imitate any of the sounds he heard. He communicated mostly by pointing. He used only 2 gestures consistently in a symbolic way: one (right thumb upwards) to mean "good" and the other (right thumb pointing horizontally towards the child's chest) to mean "me" or "Saïd". He reacted adequately to the following words only: "Saïd", "good", "bad", "yes", "no" and "more" (we assume he relied mostly on the intonation). However, he could not produce any of these words himself. It was unclear whether he could make a difference between phonemes.

One must note that Saïd lives in a multilingual environment: his mother tongue is Arabic, but his parents sometimes speak French with the children. Saïd goes to a Dutchspeaking special school, but some of his brothers go to French-speaking schools. Saïd's formal language therapy was always in Dutch, but he could not understand any of the language that were used with him.

During an object sorting test (Learning Development Aids) in which coloured printed drawings had to be classified in six categories (animals, flowers, bottles, food, houses, illuminants), several mistakes were noted: a hare was classified among the articles of food, a birthday-cake with candles was classified in the category of illuminants. These "mistakes" indicate that this kind of material was probably not adapted to the boy's cultural background and therefore, in this case, too ambiguous to be useful in the reeducation process.

It was concluded that Saïd was a globally aphasic boy with apraxia and agnosia and that his reeducation prognosis was limited, taking into account the fact that he had already recieved 3 years of formal speech-therapy at that time.

At the age of 5,6 years this child was thus not autonomous as far as language was concerned, he did not dispose of any structure of communication, he was treated as a mentally handicapped child. He understood only five spoken words and his name, he used only two symbolic gestures. We then decided to use NVCDs with him: there was one 60 minutes session each week; but there was no session when the child was ill, and this occured frequently. During each session the child worked under the direction of a neurolinguist (Lowenthal and Saerens, 1982) who used the techniques described previously (Lowenthal, 1982).

d. Follow-Up

A reexamination in December 1983 at the age of 7,7 years, • showed that the boy's condition had somewhat improved: during the last year he had been less frequently ill than before and his swallowing problems seemed to have lessened. For the first time, it was possible to examine his language comprehension by means of the Boston Diagnostic Aphasia Examination (Goodglass and Kaplan, 1972). On the auditory word-discrimination subtest, he scored 30.5/72. The highest score was found in the semantic field "verbs" (8.5/12), while the lowest score was on form name recognition (0/12). 6/12 was obtained for numbers, colours and letters. Most errors referred to objects outside the semantic field involved: instead of a circle, a key was indicated, instead of pink, and S; instead of T, somebody walking. Only 2 out of 18 names for body parts were recognised.

Comprehension of commands had developed even less: the score on the complex commands subtest was 0/15. When asked to make a fist, Saïd did not know what to do, although he could imitate the gesture when it was shown by the examiner. When asked a second time to make a fist, Saïd again looked puzzled. The comprehension for easy phrases containing spatial prepositions like up, under, in front of, ... was examined by a picture-pointing task. Although each time a picture was indicated on command, none of the answers was correct. The boy in fact just followed the sequential order in which the pictures were laid in front of him from left to right.

Spontaneous speech was absent and repetition, even of single sounds, was impossible. Sometimes, he could spontaneously produce a voluntary laryngeal sound but usually he was quiet even when laughing or crying.

He expressed himself by means of natural gestures, mostly consisting of indicatory movements, but also containing arbitrary signs. Thus the two index fingers pressed together had become to mean "is the same as". When the two index fingers were crossed, this had become to mean: "is not the same as". Both signs had been taught to the boy by the examiner.

Saïd had learned to write his name at school, but he was unable to name in writing any of the pictures (e.g. a cat, a hen, ...) shown to him. He could copy monosyllabic single words but sometimes produced mirror images for letters like S and d. He could not write from dictation.

He could understand a few words like "cheese, hand, ear, doll, hen, bus, sun" while reading the words in a picture-pointing task, but again written phrases and commands were not understood.

Auditory agnosia had improved: in an auditory recognition task, a score of 16/20 was obtained. In this test, every sound had to be matched with one of 4 drawings. Jumping was confused with climbing stairs and a vaporizer was recognized as a van. The boy was now

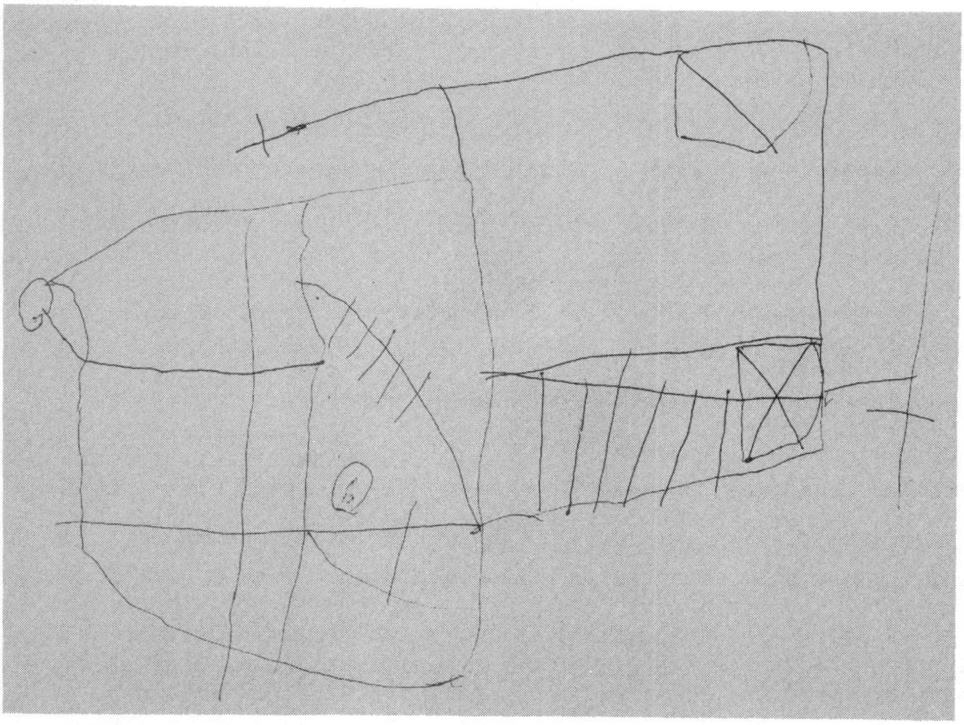

Figure 2. Saïd's reproduction of a complex REY figure (age: 7,10 years)

able to imitate simple hand positions shown to him by the examiner. His construction praxia was tested by means of the complex figure of REY (Rey, 1959) (fig. 2): the general design was distorted and many spatial errors were noted.

In February 1984, at the age of 7,9 years intelligence could be tested for the first time by means of the coloured progressive matrices of RAVEN. Former attempts at the ages of 5 and 6 years had always failed due to the boy's lack of understanding what was expected from him. A score of 12/36 was obtained (pc. 5). Many spatial errors were noted.

e. Conclusion of the Case-History

Although a positive evolution of the linguistic capacities was noted between the ages of 5,6 years and 8,6 years, Saïd still remains a severely aphasic child, unable to understand spoken and written phrases. The child still cannot speak, but he tends to use a few gestures spontaneously in a symbolic way. His behaviour shows that he now understands a few words, presented verbally.

Attempts to teach him a gesture-language and the BLISS-symbols at school have, until now, almost completely failed. The results described further will show that he learned to solve logico-mathematical problems and to express by his behaviour that he knows how to solve them.

4. MATERIAL

We used 3 different devices with this child: DIENES' Attribute blocks (Dienes and Golding, 1970), a peg board and COHORS-FRESENBORG's Dynamical Mazes (Cohors-Fresenborg, 1978). We also used a well known game: Master Mind; this last material has not been used as an NVCD.

The set of Attribute blocks (A-blocks) consists of plastic bricks; each block can be defined by 4 data: shape, colour, size, thickness. There are 4 possible shapes (rectangle, square, circle, triangle), 3 possible colours (red, blue, yellow), 2 sizes (big, small) and 2 thicknesses (thick, thin). In order to have all possible combinations one needs 48 blocks. This makes sorting and other logical exercises possible.

A peg board is a white plastic board furnished with holes. In these holes one can put coloured plastic nails. The nails or pegs are defined by two variables: their colour and the shape of the head's shape. There are seven colours: yellow, green, red, orange, pink, light blue and dark blue; the heads can be squares or quarters of a circle (we call these "triangles"). Using these coloured nails, children can make a mosaïque. But the pegs can also be used as an introduction to programming languages (Saerens, 1985).

The Dynamical Maze consists of bricks which can be placed on a base-board. These bricks constitute the basic elements needed to create a railway network: straight rails, curves, bypasses and switches. This material has built-in constraints which purposely restrict the number and the kind of combinations a child can make with the pieces: a "train" can only go through a "rail" in one direction, each piece is a "one way" element. As with the previous devices, the teacher does not need to tell the child that there are restrictions, the built-in constraints automatically impose them. They have been described in great details in LOWENTHAL and HARMEGNIES' paper (this volume). Master Mind is "a game of cunning and logic for two players" (Invicta Plastics, Ltd. 1975). The material consists of a plastic decoding board with 12 rows of large holes (code peg holes), 12 rows of small holes (keypeg holes); 5 shielded holes (for the hidden code), 2 rows of score holes and 2 peg compartments. A shield is used to hide a secret code. There are approximately 160 round headed coloured pegs, about 20 for each of the 8 colours used in the game (black, white, yellow, green, blue, grey, orange, red). There are approximately 40 small headed pegs (about 20 black ones and 20 white ones).

The decoding board is placed between the two players so that the 5 shielded holes for the secret code face the codemaker. He secretly puts 5 code pegs in 5 holes behind the shield. The codebreaker now will try to duplicate the exact colours and positions of the secret code. Each time the codebreaker places a row of code pegs (they are left in position throughout the game), the codemaker must give information using the following rule: he places as many white key pegs as the code breaker has placed code pegs of the <u>correct</u> colour in the <u>right</u> place, he also places as many black key pegs as the code breaker has placed code pegs of the <u>correct</u> colour but in <u>wrong</u> places; the number of empty holes represents the number of totally incorrect (wrong coulour) code pegs. Since Master requires no oral communication, it is an ideal game for people with hearing or speech deficiencies.

5. METHOD

In each case we first let the child play with the material, we then asked him to solve logical problems. The problems were introduced in a non-verbal way by giving examples of partial solutions, by rejecting ("no" was understood by the child) the child's movements which led to a mistake, and by approving ("yes", "good") the child's movements which led to a correct solution. We kept doing this till the child adopted an attitude which was adequate to solve the problem we had formulated. It was then possible to ask the child to solve a similar, but slightly different problem: it was no longer necessary to interfere with the child's actions. We cannot know how the child understood the problem, we can only say that we interfered until the child had adopted by trial and error a solving strategy which we hoped he would extend to similar problems. We describe here some of the problems we used and the way they were introduced.

We started using A-blocks and asked the child to sort the blocks by coulours: a large Venn-diagram was drawn on a sheet of paper, using a red pen; a red block was put inside, a blue block and a yellow block were put outside the diagram. The child was then invited (by a hand gesture) "to continue sorting the blocks": no word was used. Each time the child correctly put a block, we said "Good, Saïd", but when he made a mistake we said "No" and took away the badly placed block. The child learned a rule by trial and error.

The pegboard was firstly used to introduce sequences such as: one yellow, two reds, three blues, one yellow,... As described above the child was silently invited by hand gestures to continue the sequence. "Good, Saïd" was used when he correctly placed a peg, "No" and removal by the experimenter was used for a badly placed peg (a peg of a fourth colour, or a peg placed "out of order" in the sequence). The child had to discover a rule by trial and error.

The dynamical mazes were firstly used to let the child build paths. The experimenter showed the starting brick and the arrival brick. As previously, gestures were used to ask the child to build a path joining the two bricks. In this case several solutions are always possible. The positive and verbal reinforcement "Good, Saïd" was used as long as his solution could be extended in order to join the two given bricks. As soon as the technical constraints of the material (Lowenthal, 1984a) made it impossible to extend it to reach a complete solution, the experimenter said "No" and took away the last brick Saïd had placed on the base board. Here again the child learned to use the material by trial and error.

In order to teach Master Mind to Saïd, we introduced a first simplified version of the rules: in our version, key peg hole number 1 gave information about the corresponding code peg hole (i.e., code peg hole number 1); the same occurred for the other key and code peg holes. Moreover, a white key peg meant that the code peg placed by the codebreaker was correct (good colour and right place) while a black key peg meant that the code peg was wrong. In that way, the correct code could be found by the condebreaker after 9 trials. The child was invited by gestures to discover by trial and error the meaning of the black and white keypegs.

6. RESULTS

We will present the results concerning the three NVCDs we used and the game of Master Mind separately. The reader should keep in mind the fact that these NVCDs were in fact used more or less simultaneoulsy by the child. For each NVCD results will be presented in chronological order.

6.1. The Attribute blocks

Saïd used the 48 A-blocks for the first time in November 81; he was then 5,6 years old. The child was asked to sort the blocks: each exercise was introduced by a short training period in which some examples of correct solutions were shown. These exercises were introduced in the following way: a VENN-diagram such as that shown in figure 3 was drawn on a piece of paper. The diagram was drawn in blue, yellow or red whenever the experiment wanted to observe the subject sorting the blocks according to their colour; the diagram was drawn in black when one wanted to observe the subject while he was sorting the blocks according to another characteristic (shape, thickness or size). The experimenter introduced the problem by placing some blocks having the required characteristic into the diagram and some blocks which did not have the required characteristic outside the diagram.

During the first session (November 8,81), Saïd very swiftly learned to sort the blocks according to their colour. He was also able to find one of the 4 blocks among the 48 blocks corresponding to a drawing the experimenter made (this drawing gave only information about the shape and the colour, thickness and size were not taken in account). Saïd was also able to make a drawing correctly representing the shape and the colour of a block which was shown to him, although his apraxia made this drawing activity difficult. During the second session, the subject proved by his behaviour that he was able to learn to sort blocks according to their form. He did not succeed in sorting blocks according to size or thickness.

Symbols such as those shown on figures 4.a, 4.b and 4.c were soon introduced. They represent respectively the shape, the size (big or small) and the thickness (thick or thin) of the blocks:

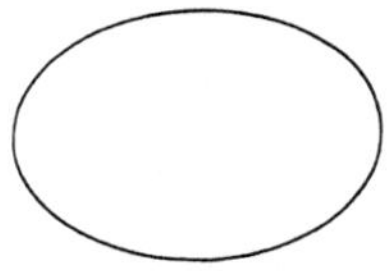

Figure 3. A VENN diagram

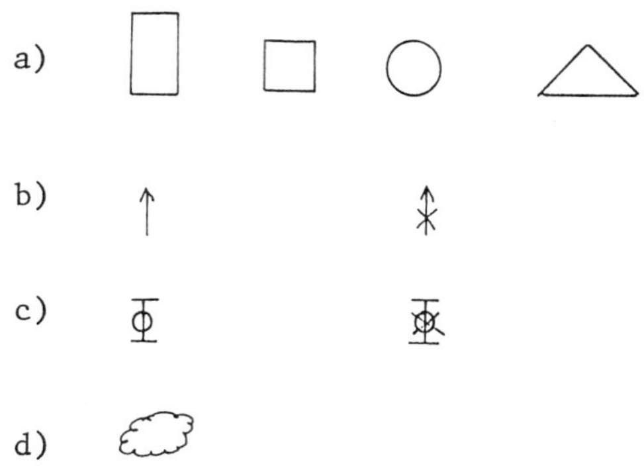

Figure 4. These symbols are used to represent a) shape, b) size and c) thickness of A-blocks. The cloud shown in figure 4d should be coloured in red, blue or yellow in order to represent the corresponding colour.

the four symbols of figure 4.a are associated to positive characteristics, while the symbol used for small (or not-big) and the symbol used for thin (or not-thick) are simply the symbols used for positive characteristics but conpleted by a cross. These symbols were introduced using techniques described by PREMACK (1976) and were drawn in black inside a VENN-diagram; the purpose of this exercise is obvious: Saīd had to put inside the diagram all the blocks having the given characteristic.

In April 1982, the use of coloured VENN-diagrams was abandoned. Three new symbols were introduced as shown on figure 4.d: a red cloud represented "RED", a blue cloud represented "BLUE" and a yellow cloud represented "YELLOW". Sequences of two symbols were then introduced.

The sequence shown on figure 5 represents "a triangle of colour 1". When the experimenter confronted the child with such a sequence, the subject had to produce one of the four blocks having these two characteristics. When the sequence was placed inside a VENN-diagram, the subject had to place all the blocks having these characteristics inside the diagram, while leaving all the other ones outside. The child showed by his behaviour that he was able to sort the blocks using two positive criteria simultaneously.

Figure 5. A triangle of colour 1.

Figure 6. A small thin triangle of colour 1

Sequences of four symbols were introduced for the first time in June 1982. The sequence shown on figure 6 describes precisely one block: the rectangle of colour 1, which is small and thin. The experimenter always presented the criteria in the same order: first shape (which reduces the possible choices to 12 blocks), then colour (which further reduces the possible choices to 4 blocks) followed by size and finally by thickness. The subject frequently confused a crossed (or negative) symbol for its positive equivalent. However when blocks were to be placed in a VENN-diagram, as described above for the November 81 sessions, the child was now able to sort the 48 blocks according to shape or colour, but also according to size and thickness.

More negative symbols were introduced in June 1982. The experimenter used coloured VENN-diagrams crossed by a black cross and shapes drawn in black and crossed by a black cross. When such a symbol was used, the child had to show any block which did not have the crossed characteristic. Using all these symbols in the context described for the November 81 sessions, the subject was able to sort all the blocks. This was not the case when he had to rely only on information such as that shown on figure 6.

At the end of July 82, Saïd was taught to draw the symbols himself and to use them to describe blocks. This proved to be difficult because of the child's kinetic apraxia. To overcome this problem the experimenter taught him to use small cards with the symbols drawn on them. By the end of August 82, the subject was able to use these cards and to "name" each characteristic of a given block; he was also able to generate a sequence of four

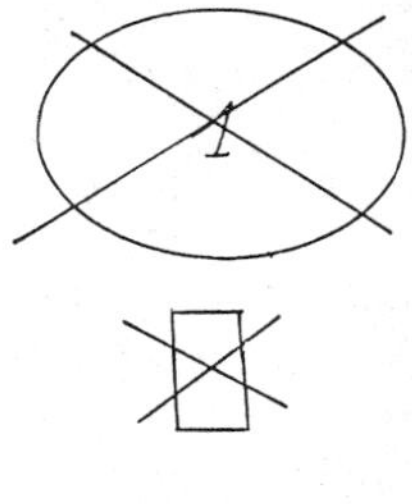

Figure 7. The upper part represents a VENN diagram drawn in colour 1 and crossed: this is the set of all blocks, except those of colour 1, the lower part represents all non-rectangles.

symbols which described exactly just one block (as shown on figure 6). The confusion between positive symbols and their negative equivalent, which was one of the problems encountered by Saïd in the decoding activity (finding the unique block corresponding to a given sequence), could also be noted in this coding activity (creation of a sequence describing a given block). The problems met in the coding activity seemed to be less important than those concerning the decoding activity. This difference between coding and decoding tallies with oberservations made by LOWENTHAL and MARCQ (1981) and concerning normal 6-year olds using another NVCD. Saïd was able in August 82 to solve 3 exercices out of 5; usually, only one of the four symbols was not correctly understood. This was mostly with the symbols for size and thickness, i.e. with the last symbols of the sequence. The crossed (and negative) symbol was still frequently mistaken for its positive equivalent.

Since September 82, after 12 hours of training with that particular NVCD, all mistakes have disappeared. Saïd is now able to sort the 48 A-blocks according to 1,2 or 3 criteria given simultaneously; he is also able to select the unique block corresponding to the 4 criteria represented by their symbols. By October 82, the last mistakes concerning the coding activity had also disappeared. Since then, one can say that Saïd has been able to use a basic vocabulary to represent concrete objects of a small universe (48 elements) and to find the unique object corresponding to a given description.

In the mean time, in November 81 the child had discovered that it was possible to cover a big rectangle exactly by three small rectangles: thus he forced the experimenter to accept as a "good answer" and element of the set of big rectangles, a collection of three small rectangles arranged in order to simulate a big rectangle (the factor colour was not taken into consideration). Very soon after he had discovered this property of the blocks he used, he covered a large thick rectangle by a large thick square and a small thick rectangle. In both cases this activity had been spontaneous and unexpected; it constituted, however, the creation of correct extended answers by the introduction of a kind of concatenation rule combined with a replacement rule. Further exercises showed that the child was indeed able to decompose big blocks into small elements and to organize a collection of small blocks in order to represent one big block.

Shortly after Saïd had spontaneously introduced these implicit combination and replacement rules, he was confronted with a "missing block" task. The experimenter placed the blue and yellow, large, thick rectangles next to one another. Since there are three large, thick rectangles in the set of A-blocks, one was missing: Saïd placed the red one next to the two others. The experimenter then placed a large, thin, yellow rectangle on top of

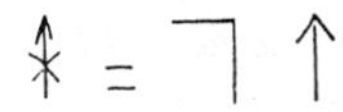

Figure 8. SMALL equals not BIG.

the thick corresponding rectangle. The subject completed this by placing the corresponding blue and red ones on top of thick rectangles of the same colour. The large, thick, yellow square was placed behind the corresponding rectangle. The child reacted by placing the two other large thick squares, and then the three thin squares on top of the thick squares of corresponding colour. Thus the child behaved as if he was able to use the criteria in order to discover some kind of pattern in the way the blocks had to be placed. This implicit but correct use of logical connectives gave us the idea to introduce logical connectives in an explicit way.

This explicit introduction occurred in December 82. These connectives were drawn in green: this differentiated them from the symbols used to represent the characteristics of the block. The equality relational symbol was introduced simultaneously and also drawn in green. We describe here how the negation operator was introduced. Until December 1982, negation was part of the symbol itself: every positive symbol shown on figure 4 could be crossed in black and thus define a negative symbol whose meaning was the negation of the meaning of the corresponding positive symbol. This made it impossible to use double negation or to represent "NOT SMALL". In December 82, the equation shown on figure 8 was shown to Saïd. According to us, this equation meant that the old symbol for "SMALL" was replaced by a sequence of two symbols: a new symbol (for "NOT") and the symbolf for "BIG". The equality symbol was also used for the first time.

The experimenter then asked the subject to find the block corresponding to sequences similar to the one in figure 9: the

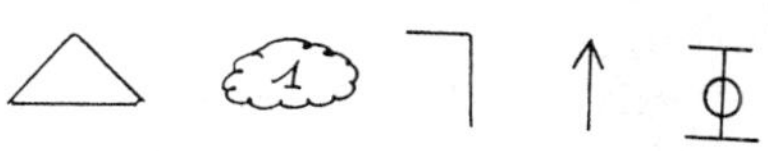

Figure 9. A small, thick rectangle of colour 1

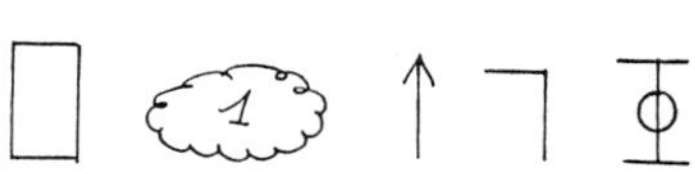

Figure 10. A big, thin rectangle of colour 1

negation operator was always immediately in front of the symbol for "BIG". After a few exercises, the child seemed to have discovered, by trial and error, the meaning of the "NOT" symbol in this context. The experimenter then introduced the sequence shown on figure 10: the "NOT" symbol was no longer immediately in front of the symbol for "BIG", but in front of the symbol for "THICK". Saïd spontaneously took the correct block. The child proved that he had perfectly understood the meaning of the new symbol when he reacted correctly to the use of this "NOT" symbol in front of any one of the symbols he was accustomed to use to describe characteristics of blocks. In January 83 Saïd learned in a similar fashion to use a symbol for "AND" as shown in figure 11.

More than one year later we again used the A-blocks: in April 84 they were used to introduce the "guessmy-rule" game. We give here the definition used by VANDEPUTTE (1976): "Lay out a short sequence of A-blocks so that all the students can see them... You should have some simple rule which you have used to generate the sequence. For example, your rule might be: red, blue, red, blue, and so on... After you have displayed four or five pieces (enough to give a clear illustration of your rule), ask the students to try to guess your rule. Any student who claims to know the rule should be allowed to place the next piece. If this is done correctly, you should praise him and ask that the rule be kept a secret. If the student guesses incorrectly, you should say: 'No, that is not the rule I was thinking of' and remove the incorrectly placed piece. It is important to realize that there are no wrong answers in this activity: a student could very well have a valid rule in mind which is different from yours. Hence, you should be careful to avoid saying: "you are wrong". A-blocks used in this way are obviously used as an NVCD; nevertheless the verbal subtelty described by VANDEPUTTE as good teacher's answers could not be used in Saïd's case.

We used the following rule: "Each block in the sequence has exactly three characteristics in common with the previous one and three other characteristics (or the same!) in common with the next block: two adjacent blocks thus differ by one characteristic exactly". Previous attempts to introduce this game (November 81 and July 82) were total failures: the boy did not seem to understand what was expected from him. One should note that a perfect knowledge of all the positive and negative characteristics is needed to solve this logical exercise.

Figure 11. A block which is rectangular and not of colour 1 and not big and thick.

The experimenter placed three blocks: first a blue, big, thick circle, then a red, big, thick circle and finally a red, small, thick circle. This was his initial sequence. The child was invited by hand gestures to place the remaining 45 blocks. His first try was a successful one: he placed as fourth block in the sequence a red, small thin circle (thickness was thus the only difference). His attempts to place a fifth block in the sequence were not as successful: first he placed a block which had only one characteristic in common with the previous one. The experimenter said "NO" and took this block away. Saïd then tried to use a block which also had only one characteristic in common with the last correct block. The child needed 4 other trials before he eventually found a block which differed exactly by one characteristic from the previous block.

Thus Saïd had to fill in 45 open places. The reader should realize that the probability of finding a correct solution at the beginning of the game is high, since many "good blocks" are available; this probability is also very high at the end of the game since there are not many possibilities left. The situation is quite different in the middle of the game since the number of available good blocks can become very small even if many other blocks remain available: the player could even play in such a way that he would arrive at a dead-end, he would no longer be albe to continue although many blocks have not yet been used. Saïd's attempts are summarized in table 1: the number of trials he needed for each case is shown in this table which also shows the evolution

Table 1.

Case	Number of trials	Case	Number of trials	Case	Number of trials
1	1	16	8	31	1
2	6	17	1	32	9
3	1	18	2	33	1
4	2	19	1	34	3
5	1	20	2	35	1
6	5	21	1	36	2
7	1	22	6	37	1
8	1	23	1	38	1
9	1	24	5	39	1
10	11	25	1	40	4
11	1	26	1	41	1
12	4	27	1	42	1
13	1	28	19	43	1
14	1	29	1	44	2
15	1	30	1	45	1

Table 2.

Number of trials	Number of cases
1	29
2	5
3	1
4	2
5	2
6	2
8	1
more than 8	3

of the game. One should note that the child made 18 unsuccessful trials to solve case 28. At this point, the experimenter gave him a solution. Saïd thus needed a total of 119 trials before all the 45 remaining blocks were correctly placed. He needed the experimenter's help in one case only (case 28). The mean trial rate was 2.64 with a standard deviation of 3.41.

It is interesting to note that in 34 cases, the child found a solution in atmost 2 trials. A qualitative analysis of these 34 cases shows that when the child succeeded in 1 or 2 trials, he usually changed thickness and never size. More detailed results are shown in table 3. One must note here that thickness is a binary criterion while shape is not binary, ... but size is also binary!!! Further studies concerning this specific problem are in process using a device such as that described in HARMEGNIES and LOWENTHAL (1984) and LOWENTHAL and HARMEGNIES (this volume).

We purposely chose to describe the child's evolution, when using A-blocks, with as many details as possible: all the other devices were introduced by means of the same method. We first let

Table 3.

	Kind of difference	Number of times
Success in one trial	Thickness	22
	Shape	5
	Colour	2
Success in two trials	Thickness	2
	Shape	2
	Colour	1

the child play with the devices, then we let him guess by trial and error what kind of problem we wanted him to solve or what kind of concept we wanted him to acquire, we finally tried to use these concepts to let the child acquire further knowledge. A detailed, step by step, description of the child's evolution with the other devices would thus not be relevant here: we will simply give the main results with the dates at which they were observed: this should enable the reader to compare the child's evolution when using other NVCDs to that described for A-blocks.

6.2. The Pegboard

We introduced the pegboard in June 82. From the beginning Said (6,1 years) was able to continue simple sequences on the pegboard. The sequences were easy: "one red, one yellow, one red, ..." or "one yellow, two reds, three blues, one yellow ...". Such sequences, involving a loop of length smaller than 4, could easily be reproduced, the longer ones could not: when more pegs were involved, Saīd usually forgot to place the last element of the loop and started the new loop too early. The child sometimes made mistakes which were corrected as usal. On June 27 he made a purposeful mistake and immediately looked at the experimenter's face. He obviously expected the usual "NO". The experimenter did not react and the child immediately put his hands in front of his face in a gesture apparently meaning "Oh! What have I done?", he then corrected his purposeful mistake and looked, laughing, at the experimenter: the child adopted an attitude which apparently meant "This was a mistake, I know it and I did it voluntarily".

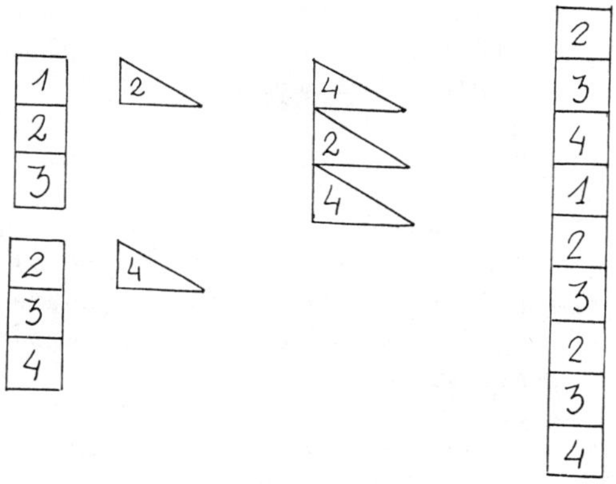

Figure 12. On the left, two subroutines with their "names" respectively triangular peg of colour 2 and triangular peg of colour 4), in the center a program consisting of three names for subroutines, on the right the execution of this program.

The pegboard was later used to introduce the first elements of programming. For these exercises we used short sequences of square headed pegs: each sequence received a name represented by a triangular peg and thus became a subroutine. A short sequence of such triangular pegs represented a program. One could use this setting to ask three kinds of questions. Firstly, if the subroutines and the program are given one can ask a child to produce a long sequence of square pegs by replacing each triangular peg of the program by its "meaning" (the child has to execute the program): i.e. the child sees on his pegboard the equivalent of the center and the left of figure 12. He must build the righthand part using square pegs. Secondly, using the subroutines and the execution of the program, one can ask a child to describe (using triangular pegs) the program which has been used: i.e. the child sees pegs on the parts of his pegboard corresponding to the left and right parts of figure 12. He must build the equivalent of the center part. Thirdly, using the program and its execution, one can ask a child to define the subroutines using pegs: i.e. the child sees pegs on the parts of his pegboard corresponding to the right and center parts of figure 12, he must reconstitute the left part.

In October 83 Saïd was confronted with this kind of exercise. He learned to handle this new kind of problem as he had done for other NVCDs: by trial and error. His learning process was similar to that described for the A-blocks. The experimenter observed several mistakes, comparable with those done by learning disabled children without aphasia (Saerens, 1985).

In December 83, Saïd was only able to solve some exercises of the first kind mostly by trial and error. These exercises were abandoned until March 84; by that time the child's performances seemed to have improved tremendously. Further studies are in progress: the same material can be used to introduce recursion and a pseudo-LOGO used to program drawings on the pegboard.

6.3. The Dynamical Mazes

The Dynamical Mazes were introduced for the first time in November 81: the child only learned how two pieces could be connected in a correct fashion, besides that the use of this material was a failure. The material was used again in August 82. The child had discovered which pieces he should use but did not yet know exactly how to orient them. In January 83, Saïd (7,1 years) had learned how to use all the pieces. The child learned to build complex paths when entry and exit were shown, to represent them by a drawing and to use a simple diagram as his only information to build a corresponding network on his baseboard (Lowenthal, 1984a). Saïd used gestures to explain that he indentified the actual network made of bricks, his own drawing and the diagram. The child consistently used the same gestures respectively for "same" and for "different". These gestures will be described later in this paper.

Saïd learned to use the bricks by trial and error and was considerably helped by the technical constraints. The mistakes he made and the difficulties he encountered are comparable with those described by LOWENTHAL and MARCQ (1980, 1982) for normal 7 years old. One must mention here that Saïd, as most normal children, first drew a track with his finger from the entrance-brick to the exit-brick, and then started to place bricks on the base board and to take care of the technical constraints.

Saïd learned to read and write small numbers at school: his knowledge of numbers is better than 10, but does not seem as good as 20. This knowledge was used in October 84 to try exercises which had been used with normal children: Saïd received a pile of cards, with one number written on each of them; the cards were in the correct order (1, 2, ...); the experimenter asked the child to make the "train" go through the maze and then to put the card "bearing the number of the train" next to the exit used by the train; the experimenter suddenly interrupted Saïd, took the maze away and invited him to continue sorting the cards. The purpose of this exercise was to examine whether the child was able to "guess the functioning rule" of the network (or mazes). The first results showed that the child really knew the difference between even and odd, but also that the child was not yet able to make long term predictions in November 84 (8,6 years).

6.4. Symbolic Gestures

We have already mentioned that the child has not been able to learn a classical sign language at school. At the age of 5,6 years he consistently used two gestures in a symbolic fashion. While using NVCDs, new gestures appeared: the child started using some of them, always in the same context, and kept using them during a short period. Sometimes the experimenter capitalized on such a gesture to create a "symbolic gesture" which the child could use to communicate. This happened when the child joined his forefingers: he used this gesture firstly to refer to his brother, later to refer to his father. The experimenter then started to use a concrete material (the A-blocks) to let the child associate to this gesture a meaning similar to "same": the fact that blocks belonged to the same category (according to one criterion) could be indicated using this gesture. To show that blocks did not belong to the same category, the child learned to use another gesture: crossed forefingers.

The child kept using gestures designating a plane or a car: these gestures are not abstractions, the child uses the enactive representation mode (two extended arms playing the role of wings for "a plane", both hands turning an imagniary steering-wheel for "car"), (Bruner, 1966). He also started to make more and more drawings: his apraxia seemed to disappear and he used simultaneously BRUNER's enactive and iconic representation modes. In June 83, Saïd (7,1 years) was able to write his name and that of the coperimenter but no more. He was also able to use and draw symbols representing characteristics of A-blocks and symbols for logical connectives. He could also draw diagrams for Dynamical mazes.

6.5. Master Mind: a Game used as an Application

Master Mind was introduced for the first time in August 84. Saïd did not know this game. We used the "easy" rules described above. The experimenter chose the following code:

red, grey, orange, green, blue.

He then invited Saïd by gestures to fill in the five holes of the first code-trial row. The child placed the following pegs:

black, green, blue, orange, grey.

None of the pegs was "of the correct colour and at the right place": the experimenter placed five black key pegs. As the child did not know at that moment what black key pegs meant, the experimenter pointed with one hand to each trial-code peg while pointing to the floor with the thumb of the other hand: this gesture meant for Saïd that the code peg was "bad" or "wrong". Saïd was urged to try again, and obtained the same answer till he produced the following trial-code on the third row:

black, white, orange, blue, yellow.

The experimenter pointed to the black code peg, placed a black key peg and made the "thumb down" sign. He repeated this for the white code peg; but he put a white key peg while pointing to the orange code peg and made a "thumb up" sign. For the remaining code other pegs he proceeded as for the black one. He immediately inserted an orange peg in the correct hole of row 4. Saïd completed this row in the following way:

grey, white, orange, green, grey.

The "thumb up" and "thumb down" procedure was repeated and the key pegs were placed by the experimenter. Saïd immediately placed an orange peg and a green peg in the correct place in the next row. He worked by trial and error for the other pegs and found the solution in row 12 only.

In a second exercise, Saïd was the codemaker while the coperimenter was the codebreaker. While the boy used only the thumb up and down procedures, the experimenter introduced black and white key pegs according to the information given by Saïd. Saïd's behaviour showed that he understood the meaning of the game and could play it adequately. The coperimenter was able to find the code after only 5 trials.

Saïd, as codebreaker, did not seem to understand the meaning of the black key peg: he frequently reintroduced code pegs in places where he had already used them before. Further games always making use of the combination of thumb-procedure and key pegs, were played. After 5 turns, most errors had disappeared, and Saïd was able to find the solution after 8 to 9 trials.

New rules were then introduced: the new rules give a better approximation of the actual rules but the position occupied by a key peg continues to give information: a black key peg means that the colour of the corresponding code-trial peg is present in the hidden code, but not in the corresponding place; a white key peg means that the coresponding code-trial peg has the correct colour and occupies the correct place: an empty space in the key peg row means that the colour of the corresponding code-trial peg is not used in the hidden code. This new meaning of key pegs was explained during a game where the "hidden code" could be seen by Saïd, who was the codebreaker. The experimenter used hand movements to show that code pegs corresponding to white key pegs were perfect but that those corresponding to black key pegs had to be used in another place.

The experimenter then asked Saïd to play a real game (with a hidden code): the child's behaviour showed that he had not understood the new rules, and continued to play the game using the previous convention. The experimenter decided to use the VENN-diagrams which had been used for A-blocks: a diagram was drawn next to the Master Mind board and pegs of the appropriate colour where placed in it ("good colour") or out of it ("colour not used in the code") as soon as this information could be obtained. The following code was used:

orange, blue, black, grey, red.

And Saïd tried:

orange, grey, red, white, green.

The experimenter gave the following information using key pegs:

white, black, black,

He also placed orange, grey and red pegs inside the VENN-diagram, but white and green ones outside. Saïd needed time and help to understand the new convention; he was able to play without help after three more exercises. The "thumb up" and "thumb down" signs were abandoned: the key pegs and the VENN-diagram serve as the only guides.

Saïd learned in the same fashion to play this game as codemaker: in November 84 he was able to give adequate responses, using the key pegs, in such a way that his opponent could solve the game.

He is now able to play this game, which he likes very much, with his parents, his brothers and his sisters.

6.6. Further Results

One must also mention that the child has a watch and can use it adequately. He is also able to write a few words (the first one was his name), to read and write numbers, at least numbers smaller than 12. The child also knows the difference between even and odd numbers. At school, he has been taught some notions about basic arithmetical facts, but it has been impossible until now to evaluate his actual knowledge in this field.

All these notions have been acquired one way or another, using some kind of communication system which is not necessarily a structured one. As this child is a living human being, as he has a family and goes to school, it is impossible to establish where exactly the child acquired these notions: the role of NVCDs in this acquisition cannot be correctly evaluated.

7. DISCUSSION

When we examined the child for the first time he was 5,6 years old and did not seem able to communicate in a structured way with his relatives or other people; he was not able to produce a solution for logical exercises classically used as intelligence tests. He had not been able to learn any substitute for the usual articulate and verbal language. Three years later the situation has changed: the child is still unable to speak but he understands a few words; he is able to produce solutions for difficult logical exercises and he is able to communicate, although non-verbally, in a structured way using gestures and symbols he has often introduced himself. The "symbolic" meaning of such gestures can only be defined in terms of "child consistent behaviour". Many things happened during these three years: Saïd started to go to primary school, he grew older and lived with parents who care a lot for him; he also started using NVCDs (Saïd had attended Kindergarten, NVCDs were in fact the only new things in his life). All these events could have influenced, one way or another, the evolution of the child, and some of these possible influences could not be controlled: this makes a classical discussion of the evolution of the communicative ability in this child meaningless. This is why we chose to formulate six precise questions which can be answered on the basis of the evidence we have.

1) Is the subject able to use concepts and to acquire new ones on the basis of those he is already able to use?
2) Is the subject able to use operations, to apply logical connectives to the concepts he has mastered and to learn new operations?
3) Is the subject able to associate symbols to the concepts and to the operations he uses?
4) Is the subject able to use and acquire combination rules

and to use them when symbols and/or objects are manipulated?
5) Is the subject able to interpret a complex sequence of symbols and/or objects built using such rules?
6) Is the subject able to show that he is conscious of what he is doing? More precisely does the subject show what his intentions are when he is using symbols and/or objects in a rule governed fashion?

In fact the first four questions could be summarized as follows: is the subject able to learn a type of propositional calculus; question 4 obviously refers to what KLEENE (1952) calls formation rules for formulae: "Certain subcategories of the formal expressions, by definition analogous to the rules of syntax in grammar". Question 5 concerns the ability the child has to create a model corresponding to a well formed sequence of "signs" (symbols or concrete objects used as symbols) and to adapt his behaviour to this model (i.e. creating a problem solving strategy). Question 6 concerns among other things, the child's ability to use the framework of well formed sequences to lie or to make purposeful mistakes which correspond to jokes. In fact this question concerns the pragmatics of Saïd's communication, i.e. "the intention of the speaker during communication or the functions that language serves, as well as how one modifies language according to contextual situations" (Aran and Nation, 1982).

We voluntarily restricted this discussion of our observations, to the logical domain: NVCDs have been built to observe and measure logical activities in children. It might also be argued that the use of NVCDs made it possible to favour their communicative development; but although we believe that this is the case, we are not going to discuss this point since no clear-cut definition of "communication" has been introduced: this is not the case for "formal logic".

7.1. Concepts

Saïd's way of using A-blocks and of sorting them clearly shows that he was able to use "obvious" concepts such as "shape" and "colour" in November 81; but also that he was able to acquire the concepts "thickness" and "size" very rapidly. It is also important to note here, as evidence for a positive answer to question 1, that Saïd quickly learned to sort A-blocks according to one, two, three or four positive or negative characteristics. The child was also able to learn and use the size and colour criteria when he manipulated the pegboard. Similarly, he was able to sort and to use the bricks used for the Dynamical Mazes adequately. Finally the child learned to use the concept code peg and key peg when he learned to play Master Mind. All this leads us to answer "YES" to the first question.

7.2. Operations

Simple operations were implicitly used either when a symbol was crossed in order to introduce "negation", or when a concatenation of symbols chosen among those shown in figure 4 introduced implicitly an "AND". Moreover the child learned to read a sequence of triangles similar to that shown in figure 12 as an "AND THEN". All this pleads in favor of a "YES" answer to question 2.

7.3. Symbols

The subject obviously learned to use arbitrary symbols when using the A-blocks. The differences we observed between shape and colour versus size and thickness might be due to the fact that the first two symbols are iconic and the two lost ones symbolic (Bruner, 1966). The first use of the pegboard introduced a number concept and a concept of subsequence; this number concept was later associated to symbols which the child did not learn with us: this proves that he was able to use symbols by himself or at least in a school setting, while this was not the case before NVCDs were introduced (no evidence of number symbols before the child was 8 years old). The second use of the pegboard introduced another instance of the concept of subsequence; these subsequences have a name (the triangular pegs) which are objects used as symbols. More evidence of symbolic activity can be found in the associations of diagrams with Dynamical Mazes, and of parts of a diagram with specific bricks. The consistent and adequate use of gestures and their association with non obvious meanings ("two joined forefingers" is used as "same") is a last example of acquired symbol. All this forces us to give a positive answer to question 3.

7.4. Combination rules

The subject spontaneously used a replacement and a concatenation rule when he exactly covered a big rectangle with smaller A-blocks. His behaviour when looking for the "missing blocks" is another example of adequate use of combination rules. He never used a sequence of symbols (shown in figure 4) which contained more than one symbol per category: he apparently considered such sequences as illegal. His way of using the pegboard (either to continue sequences or to use programs) show that the subject is able to learn to use a combination rule. This seems to indicate that the child can apply a combination rule.

7.5. Interpretation of Sequences

Here the situation is not the same. The child is obviously able to interpret a simple sequence coding one A-block, or a set of A-blocks. He is able to use the sequence of information he receives when playing Master Mind as code-breaker. But there is no evidence that he can interpret sequences of pegs in order to solve programming exercises of kind II or III (find the program, find the subroutines). He is not yet able to use the Dynamical Mazes and to discover the rule which governs the passages of the "trains", although he has as information a sequence of numbers and showing the order in which each exit is used. The strategy Saīd adopted when playing the one-difference game with A-blocks does not enable us to draw a conclusion with now. Further studies concerning this specific point are needed: they can easily be carried out using the "guess my rule" game and the examination of predictions concerning the Dynamical Mazes; we have indeed for these two kinds of problems a great number of data concerning normal children.

7.6. Intentions

Such a domain cannot be precisely measured by means of NVCDs; one can only give clinical (and partially subjective) observations. At the age of 5,6 years Saīd was a calm and serious boy; he was kind, was well-treated by his family and was more or less left out in Kindergarten. As soon as we started using NVCDs with him, he showed by his behaviour that he was interested in what we were doing with him: he was smiling and active. He soon started to use enactive representations in what we consider to be attempts to describe his everyday life, what happened at home, ... He also started using the same gestures consistently and in the same context: the child initiated this process.

His decision to introduce a concatenation rule and a replacement rule for A-blocks might be due to his misunderstanding of the task; it might also be a first joke about the material, a joke which was expressed through this material. There were other cases were Saīd's attitude obviously showed that he tried to trap the experimenter, especially when he used the pegboard. All this might be evidence of an important pragmatic activity which started developing after the introduction of NVCDs.

8. CONCLUSION

One can not yet claim that this child has a complete mastery of some kind of propositional calculus: we do not yet have enough evidence; it also seems probable that we will have the last threads of evidence we need in a near future. It will then become possible to add to this the information we have about some kind of pragmatic activity in the child: usually, natural gestures serve as "meta-language" when Saīd is making jokes about or with the material he is using. Does he master some kind of communication system which he uses as non-verbal language? If we follow CHOMSKY (1957) we might be tempted to say yes. If we use OSGOOD's definition of language (1980) we must say that Saīd <u>masters</u> a non human (because non-verbal) language.

What role did the NVCDs play in the child's evolution? This question cannot be answered; but we can say that Saïd's potentialities had not been completely studied and that they could only be completely evaluated using the logical and simplified settings imposed by the structure of the NVCDs. The NVCDs thus played an important part: they gave the first positive information about this child. Moreover the way in which Saïd reacted to the NVCDs seem to illustrate the hypothesis formulated by LOWENTHAL about the mental processes involved when using NVCDs (this volume): they seem to have been useful tools similar to those suggested by JOHNSON and MYKLEBUST (1967) in other situations.

The NVCDs used have thus probably favoured the child's development. But all these ideas are pure speculations: it seems better to conclude with facts.

At the start Saïd was not autonomous. He was only communicating by pointing. According to RENARD's terminology (1985), Saïd could not define his own law.

The technique which was used was easily accessible to Saïd: each new problem was a new challenge since it was "just out of immediate reach but not too far to be attained by an extended arm". This method - the use of NVCDs - is progressive and challenging: it was used to introduce prerequisites for sorting processes and other logical operations. These are the bases needed for structuring. It is not necessary to stress here the importance of sorting processes in language acquisition.

Saïd has made important progress and has started to use the devices provided to him in a more and more autonomous way.

During his lecture, at the SGAV 6 conference, RENARD said that "Autonomy results from learning to use basic elements": this can be done using NVCDs. He also said that one should educate the ear, the rest will follow: in Saïd's case, the brain lesions were and are such that the education of the ear appeared useless; so we tried some other kind of sensorial education (vision and touch, which provide the brain with much information). This seems to be a fruitful approach.

REFERENCES

Apostel, L., and Vandamme, F., 1975, Formele logika Deel 1, Klassieke systemen. De Sikkel, Malle, 1975.
Aran, D.M. and Nation, J.E., 1982, Child language disorders, Mosby Company, St. Louis.
Arbib, M., and Caplan, D., 1979, Neurolinguistics must be computational, behavioral and Brain Sciences, 2, 449-483.

Arbib, M., Caplan, D., and Marshall, J., eds., 1982, Neural models of language processes, Academic Press, New York.

Bruner, J.S., 1966, On cognitive growth, in Studies in cognitive growth, Bruner J.S., Greenfield P.M. and Olver R.R., eds., New York, John Wiley, 1-67.

Chomsky, N., 1957, Syntactic structures. The Hague, Mouton.

Chomsky, N., 1968, Language and mind. New York: Harcourt, Brace and Jovanovich.

Cohors-Fresenborg, E., 1978, Learning problem solving by developing automata networks, Revue de Phonétique Appliquée, 46/47, 93-99.

Denckla, M., 1979, Childhood learning disabilities, in Heilman, K. and Valenstein, E., Clinical neuropsychology), Oxford University Press, 535-573.

Dienes, Z.P. and Golding, E.W., 1970, Les premiers pas en mathématique: logique et jeux logiques, OCDL, Paris.

Goodglass, H. and Kaplan, E., 1972, The assessment of aphasia and related disorders, Léa and Febiger.

Harmegnies, B. and Lowenthal, F., 1984, Dispositifs de communication non verbale et ordinateurs, Humansky bernctik, 25, 318, 115-124.

Johnson, D.J. and Myklebust, H.R., 1967, Learning Disabilities, Educational Principles and Practices, Gruner and Stratton, New York.

Kleene, S.C., 1952, Introduction to Metamathematics, Amsterdam-London, North-Holland.

Lowenthal, F., 1982, Examples of auxiliary formalisms to help the development of children's logical thinking, in Language and Language Acquisition, eds., Lowenthal F., Vandamme, F. and Cordier, J., Plenum Press, New York, 113-121.

Lowenthal, F., 1984a, Productions langagières d'enfants manipulant un dispositif non verbal de communication, Revue de Phonétique Appliquée, 69, 11-46;

Lowenthal, F., 1984b, Language and mathematics - interactions between their developments, Laboratoire NVCD, University of Mons, Mons.

Lowenthal, F., and Marcq, J., 1980, Dynamical mazes used to favour communication among 7-8 year olds , in Proceedings of the 4th International Conference for the P.M.E., ed. R. Karplus, Berkclay, 370-376.

Lowenthal, F. and Marcq, J., 1981, Logic, Auxiliary Formalism and Geometry by Telephone Call, in: Proceedings of the 5th conference of the international group for psychology of mathematics education, vol. I, eds., C. Comiti and G. Vergnaud. Laboratoire IMAG, Grenoble, 1981, 265-270 , en collaboration avec J. Marcq.

Lowenthal, F., and Marcq, J., 1982, how do children discover strategies (at the age of 7)?, in: Proceedings of the 6th conference, Psychology of Mathematics Education, ed., A. Vermandel, Univers, Faire Instelling Antwerpen, Antwerpen, 1982, 287-292, (in collaboration avec J. Marcq).

Lowenthal, F., and Saerens, J., 1982, Utilisation de formalismes logiques pour l'examen d'enfants aphasiques, Acta Neurologica Belgica, 1982, 82, 215-223.

Osgood, C.E., 1980, Lectures on language performance, New York, Springer-Verlag.

Premack, D., 1976, Intelligence in ape and man, John Wiley, Hillsdale.

Renard, R., 1985, Structuro-globalism and autonomy, Revue de Phonétique Appliquée (à paraître).

Rey, A., 1959, Manuel: Test de copie et de reproduction de mémoire de figures géometriques complexes, eds. du Centre de Psychologie Appliquée, Paris.

Saerens, J., 1985, Semantic components and syntactic sequences introduced by means of multicoloured plastic pegs, Communication and Cognition, 18, 1, 24-28.

Siegel, L.S., 1982, The discrepancy between cognitive and linguistic abilities in the young child, in, Language and Language Acquisition, eds., Lowenthal, F., Vandamme, F. and Cordier, S., Plenum Press, New York, 31-38.

Vandeputte, C., 1976, C.S.M.P. Mathematics for Kindergarten, Teacher's guide, CEMREL, St. Louis.

Van Dongen, H., and Loonen, M., 1976, Neurological factors related to prognosis of acquired aphasia in childhood, in, Recovery in aphasics, Lebrun, Y. and Hoops, R., Swets and Zeitlinger, Lisse, Neurolinguistics, 4, 210-215.

Van Hout, A., and Seron, X., 1983, L'aphasie de l'enfant, ed. Pierre Mardaga, Bruxelles.

Woods, B., 1982, Acquired childhood aphasia, in: The neurology of aphasia, Kirschner, H. and Freemon, F., Swets and Zeithlinger, Lisse, Neurolinguitics, 12, 163-172.

SYNTHESIS AND FUTURE PERSPECTIVES

SYNTHESIS AND FUTURE PERSPECTIVES

F. Lowenthal

Laboratoire NVCD

University of Mons

During the 1983 conference many papers were presented (they can be found in the conference proceedings), but the discussion sessions cannot be found anywhere. Therefore, we thought that it would be worthwhile to summarize them here. The organizers purposely chose a program structure in which it was possible to rediscuss a paper presented on the first day of the conference on the final day. Such a program structure favours discussion, even if it makes an independent presentation of each discussion session impossible. We will thus make a global synthesis.

This synthesis will be my personal synthesis: I claim full responsibility for all the remarks and suggestions I make. It will be divided into three parts: firstly a few general remarks concerning what has been said, then more remarks concerning specific topics which were mentioned during the conference and my own personal reactions, the ideas which came to my mind during the discussion but which do not necessarily represent the opinion of all the participants, and finally a list of problems which seem important to me and for which I feel that the Language and language acquisition group must try to find a solution for the 1986 conference.

This text will correspond to an address delivered during the closing session in 1983. The papers contained in this book were not written at that time: the reader can compare the following synthesis with these papers and draw his own conclusions about the present state of these problems.

When Dominicy gave his conclusions at the end of the 1980 conference, he remarked that one name had hardly been mentioned during the conference: that of Piaget, and another one had not been mentioned at all: that of Chomsky. Does this indicate a trend? This time both names have been mentioned several times: there were even discussions about a "Chomskian baby". Other names of well known researchers were also used during the discussions: Bruner was quoted frequently and Bernstein and Halliday, who had not been quoted in 1980, were quoted even more frequently. The word "device" also came into the discussions very often. These are the main general remarks concerning the differences between this conference and the previous ones.

What were the new ideas which appeared during these discussions? There were interesting oppositions. On the first day Spoelders gave one definition of Pragmatics in terms of relations between speaker, listener and utterance; but Wells, who spoke immediately after him, gave another definition; according to him Pragmatics is language in use. At the end of the 1983 conference on Pragmatics and Education, we still do not know what the correct definition is: we wonder whether Pragmatics is not simply the purposeful use of Semantics, i.e. Semantics in action. A second opposition appeared when Wells presented the verbal components of language: he insisted a lot on the importance of verbal interaction. Many participants seemed, during later discussions, to share his point of view. However just as many participants insisted upon the importance of the non-verbal components of language. These non-verbal components include gestures (voluntary or not, conscious or not) but also mimics, eye movements and all the paralinguistic elements: prosodic elements cannot be neglected. This implies that one should study intonation, rhythm and music. Why should we avoid music as a study subject in a group such as this one? Or art? We could include then, of course, but we have to be careful since researchers such as Osgood consider that the "language of music" or the "language of art" are not languages at all! But he does not say anything about mime, or dance. Some Asian dances convey messages; they have a meaning and they have to obey very strict rules. Do they constitute a language? This is a first topic about which further reflection and discussion is needed.

As usal, the participants were specialists belonging to many different research fields. Nevertheless, two new disciplines were discussed. Pinxten introduced a long discussion about the interactions between Topology and Linguistics, when he described a program to teach geometry to Navajo Indians. Pinxten also insisted on the importance of tradition in language acquisition. Micronesians do not use the same kind of language as Papuans, their traditions are based on totally different contexts; what about Sapir-Whorf's hypothesis? The exact role of tradition in language acquisition might be another theme (for reflection). Tradition can be lived in very specific fashions: decorations such as masks might be used. In Belgium and in other European countries, masks are used while preparing the carnivals, and during the feast itself they are used according to a very precise ritual. Do these masks convey a message? If the answer is "yes", we cannot neglect this important element of some peoples' tradition.

There was also much discussion concerning logico-mathematical structures and a long discussion during which participants spoke about representations: cognitive monitoring was mentioned. But, should representations be structured or not? Is there any interaction between logico-mathematical structures and the kind of representations which are relevant for our group?

One word kept reappearing in many different contexts: schematization. Should schematization be formalized, or not? Is it even possible formalize it? One participant claimed that schematization cannot be studied without argumentation. In 1980, a whole session was devoted to argumentation, but in 1983 we did not speak a lot about it. The need for argumentation, and its importance when discussing schematization is another possible theme for reflection for our group.

Non-verbal communication devices were discussed at great lenght. A whole session was devoted to them and this concept was mentioned again during discussions concerning apparently totally different subjects such as Engels' paper. The discussion concerned the relevance of such a name. Important problems were raised: to what extent and in which contexts do NVCDs have something to do with communication? Are NVCDs restricted tot the field of cognitive activities and to the observation of a Piagetian "decalage"? Could one say that NVCDs can be used to favour the development of meta-communication? Can NVCDs be used to overcome limitations in the child's ability to communicate - limitations which impeded research lead by Piaget and his followers? Could this give us further information about the Piagetian stages? Some participants feel that they are not always valid; they suggest that the non-verbal domain can be used to build knowledge as well as to perceive it. Other questions arose; can children transfer the knowledge they acquired in the specific field of NVCDs to other fields? Can tools be used as open symbols?

NVCDs were used with handicapped as well as normal children; they were used for the examination and treatment of acquired specific concepts and to let children discover the notion of algorithm. But the evidence introduced in 1983 did not appear sufficient for the participants, they did not reach a common and unique conclusion about the relevance of NVCDs. We have tried to introduce more data in this book, but we think nevertheless that NVCDs constitute another reflection theme.

In connection with NVCDs the word 'drill' has been mentioned. By the way, do we have a precise definition of what is, and what is not, a drill in language acquisition? Isn't drill a form of education or treatment? or conversely shouldn't we say that education implies learning. This word has been used often during the conference, but do we really know what it means? In fact, we are often using words which seem to be basic words, and we are convinced that by doing this we can build nice models and describe nice theories; but there is a problem: the relations between speaker and listener do not always function properly since they do not always use the same basic definitions. Can this problem be avoided? It seems worthwile to think about it: the role of language in education could also be a theme for refection.

If we think about language _in_ education, we should also think about schools. Are they useful? Some participants seemed to say that school, as far as language acquisition is concerned, might be dangerous. Other participants seemed to claim that schools appear to be useless. It might be interesting to reflect about the education _of_ language.

Many other subjects were discussed during the 1983 conference, such as: story telling, communicative and linguistic competence, meaning,... But do we have a clear definition of "meaning"? The answer is no; how can people expect researchers to agree about conclusions if they do not agree about the meaning of basic terms: the situation is much easier in mathematics where these difinitions exist. Would it be impossible to create such definitions four our field of study?

A very small word was used very often. In fact it is a prefix: "meta". It was used frequently with-language but more frequently with-cognition. Here again people do not always agree about its meaning. In 1980 already Osser suggested metacognition should become a further research area: this should not be forgotten.

There were several presentations and many discussions about reading and writing. In 1980, a whole session was devoted to comprehension and production; there was no similar session in 1983: it might be interesting to study comprehension and production in connection with reading and writing problems. Anyway we should think about the following question: to what extent do we understand what a child understands or _intends_ to utter?

Bilingualism was discussed by several participants. There was a discussion about acquiring two languages simultaneously; however most researchers are able to use several languages, but they de not use them in the same context: one person can often use very different languages but each of them is used in a very specific context. This brings us back to another suggestion we made at the end of the 1980 conference: it might be worthwhile broadening the scope of classical Psycholinguistics by also studying the relations between context and language. It was suggested, in 1980, to study what we then called "Contextual Linguistics". We defined this new research field as being neither a part of linguistics, nor a restricted form of psycholinguistics, but on the contrary an expanded version which included psycholinguistics, with all its structures, but which is less rigid: formalization will be used as a technique to go from concrete to abstraction, but also from abstract to concretization".

Context has been mentioned during the 1983 conference, but my first general conclusion, is that I feel we did not look enough at context and at all its relations with actual language acquisition.

We now wish to make a few suggestions about things which have not been mentioned during the 1983 conference, or at least not sufficiently discussed.

Firstly, it seems that it is essential for our group to gather data: we do not have enough. Sometimes we present models: they look nice and interesting, but they are based upon a few observations, not enough pure facts and we cannot thus discuss their generalizibility. Secondly, we will mention here a few domains in which data should be gathered urgently: this seems to be a top priority! Data concerning the language used by infants are needed before we try to elaborate further theories; but observations concerning the mentally-retarted are at least as important since, according to Rondal, their development parallels that of normal children, but is slower. We should not forget to complete our collection of basic data by looking at all acquired speech disorders.

This brings us to another domain where information is lacking: linguists, psychologists and others do not know enough about human physiology. There is an even more important gap: we reed more information about neurological factors which could influence language acquisition or therapy concerning speech disorders, we need more clinical observations of patients; we speak about people, about what they say, about the way they say it, but we do not have the basic information.

Data should also be gathered in another domain: some of us study children's strategies during their cognitive and communicative development. It seems that we should gather more information about these strategies. This means that we should also look at early strategies, the strategies used by young infants and concerning the development of language and thought.

This brings us immediately to a third main suggestion: all the previous suggestions might sound very interesting, but in fact they cannot be put into practice immediately because of the tremendous methodological problems we have to solve before we can do any decent work. We do not mean here that these methodological problems are unsolvable: but they do exist and we think that it is important to tackle them and to try to solve them. Some of the papers in this book already describe devices which can be used to solve some of these problems, in connection with research in our field. These attempts to solve methodological problems should lead to the construction of devices like the one described by Lowenthal and Harmegnies in this book; this might be a good method of approach for another main problem: how can we approach the matter from the inside? How can we understand how a child, or anybody else we are studying, is reacting to our questions, our remarks, the simple fact that we are observing him? It might be useful to use simple logical structures in connection with this: such structures restrict the number of possible interpretations (and misunderstandings), but it might also be a very poor method since logical structures will never enable us to use a very rich and subtle language.

Other domains should not be forgotten. Why have we forgotten in 1983 to look at playing in connection with socialization? Many researchers have already looked at one of these two subjects, but not at both: it seems that not enough people have looked at infants' playing in connection with socialization, and in connection with the creation or the appearance of child's first words. This leads us to the observation of toys, of tools which as Vandamme said, can be manipulated and used as symbols. Symbols are important in the exact sciences: the progress of these domains of knowledge would be impossible without symbols. Do we really know enough about the role of real, visible, concrete objects used as symbols in the development of language? This does not seem to be the case.

There were also some discussions about computer languages and programming. This was probably not sufficient: we must learn more about these strange things since we will certainly not be able to avoid them. Those wishing to avoid computer languages but also wishing to keep working in the field of language acquisition will fall into some kind of trap: computer languages are everywhere, it is impossible to keep organizing meetings such as this one and refuse to discuss Logo or interactive programs.

The last theme for reflection we want to mention here concerns workability: are all the models which we create and describe, worable models? What does it actually mean, for a model to be workable? Does it only mean that it is accepted by a majority of researchers?

This synthesis should now come to an end, and thinking about everything we have discussed in connection with speech, writing and reading, we want to quote a last passage by an unknown author:

"Nothing, not even a good teacher, not even a good school, has ever succeeded in preventing a child who really wanted to, from learning to read". We might add: "Nothing has ever succeeded in preventing a child - be it a Chomskian baby or a Piagetian infant - who really wanted to, from learning to speak". In fact it seems that this is true for many other things which occur although we do not exactly know why and how they occur.

INDEX